ICCS 2007

Babak Akhgar
Editor

ICCS 2007

Proceedings of the 15th International Workshops on Conceptual Structures

Babak Akhgar
Sheffield Hallam University, School of Computing and Management Sciences, UK

British Library Cataloguing in Publication Data
A catalogue record for this book is available from the British Library

ISBN 978-1-84628-990-3 ISBN 978-1-84628-992-7 (eBook)

Printed on acid-free paper

9 8 7 6 5 4 3 2 1

Springer Science+Business Media
springer.com

Editorial Board

International Program Committee

Foreword

It gives me great pleasure to introduce this collection of papers to be presented at the Workshop and Industry day of 15th International Conference on Conceptual Structures (ICCS 2007) at Sheffield Hallam University, Sheffield, UK.

At ICCS 2007, some of the world's best minds in information technology, arts, humanities and social science will converge to explore novel ways that information and communications technology (ICT) can augment human intelligence, the longstanding objective of research and development efforts dating back to the pioneers Vannevar Bush and Douglas Engelbart. The objective driving research and development on "Conceptual Structures" is to harmonise the uniquely human ways of apprehending the world, with the power of computational information management and reasoning. Arising originally out of the work of IBM in Conceptual Graphs, and developed by learned researchers and business organisations such as Boeing and Microsoft, over the years ICCS has broadened its scope to include a wider range of theories and practices. Amongst these are Formal Concept Analysis, Description Logics, the Semantic Web, the Pragmatic Web, Ontologies, Multi-agent Systems, Concept Mapping, and more.

On behalf of the program committee I would like to thank all those who submitted papers for consideration. Each submission was evaluated by at least two referees. The overall acceptance rate was about 34%. I am very grateful to many colleagues who helped in organising the workshops. In particular, I would like to thank the following:

MOSAICA EU IST project for sponsoring a Workshop on Semantic and Ontology Engineering.

MATCH EU IST project for sponsoring a Workshop on Bio-Informatics.

Members of the ICCS 2007 program committee.

Last but not least Springer UK for their support.

Professor Babak Akhgar
Chair of ICCS Workshops

Contents

Section 1: Trust in E-systems and the Grid Workshop

Section 2: Quantitative Approaches for Knowledge Discovery and Decision Support in the Post-Genomic Era

Section 3: International Workshop on Rough Sets and Data Mining

Section 4: International Workshop on Ubiquitous and Collaborative Computing

Section 1

Trust in E-systems and the Grid Workshop

A Contextualised Trust Model for Distributed Open Systems

Nardine Z. Osman

The University of Edinburgh, School of Informatics, N.Osman@sms.ed.ac.uk

Abstract. This paper presents a contextualised model of trust, built upon existing trust mechanisms, as opposed to the traditional rigid models. The contextualised model is based on the view that there is no one trust mechanism suitable for all scenarios. What might be suitable for one scenario could be inappropriate for another. We bring this model to a practical level by showing that agents are capable of specifying and verifying these dynamic models at runtime.

1 Introduction

Despite the extensive research, trust remains a fundamental challenge for distributed open systems, such as multiagent systems (MAS). For instance, until now, there has been no consensus yet on what exactly trust is. This is due to the fact that trust may be addressed at different levels. On the low system level, trust may be associated with network security, such as authentication, content privacy, etc. On higher levels, the focus is on trusting entities (human users, software agents, etc.) to perform actions as requested, provide correct information, etc.

The problem, we believe, is neither in the broad spectrum of these trust issues, nor in the diversity of methods addressing them. The problem, as we view it, lies in the disability of deciding at run-time which issues should be currently addressed and how to address them. We argue that different environments and different scenarios require different strategies for dealing with trust. We then leave it to the agent to decide which strategy is most suitable for the current interaction with the given set of collaborating agents. For example, while it may be preferred to distrust new entrants (with no reputation) to the system under critical circumstances, it may be favoured to trust anyone with no explicit negative reputation under less-critical circumstances.

The agent's reasoning layer that selects the best strategy for a given scenario is outside the scope of this paper. Our work concentrates on providing this layer with the formal methods for specifying and verifying these strategies, which in effect aids its decision process. This paper mainly focuses on the proposed contextualised trust model. It starts with a brief introduction to MAS models in Section 2, followed by a presentation of the contextualised trust model in Section 3. Overviews of the specification and verification processes are presented in Sections 4 and 5, respectively.

2 Multiagent System Models

We view MAS systems as a collection of autonomous agents engaged in various interactions. Figure 1 provides such an example. In such dynamic systems, agents may join or leave at anytime. Moreover, the agents' beliefs, desires, goals, and plans are continuously changing. We believe that in such open and distributed systems the interaction groups should be created dynamically and automatically by the agents. Therefore, there should be no higher layer for coordination and control, including trust control. As a result, we split the MAS model into these two layers: the interaction layer and the agents' layer.

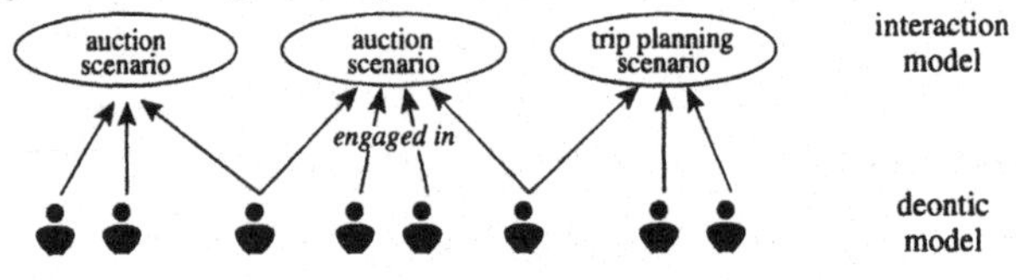

Fig. 1. MAS model

The interaction model simply states how the interaction may be carried out. Figure 2 provides a sample interaction model for an auction scenario. The interaction model is independent of the agents engaged in it. Any constraint imposed by any of the involved agents is then modelled on the local deontic level. These are usually the agents' permissions, prohibitions, and obligations. In this paper, we focus on the agent's trust constraints. Figure 3 provides a sample of trust constraints imposed by the agents on the given auction scenario.

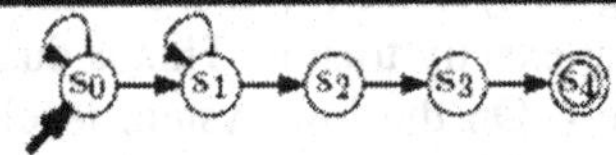

The interaction starts at state s_0, with the auctioneer sending invites to bidders. After all invites are sent, the interaction moves to state s_1 and the auctioneer then starts collecting bids. After all bids are collected, interaction moves to state s_2 and the auctioneer sends a message to the winning bidder informing it of the price won at, moving the interaction to state s_3. Finally, the bidder sends its payment to the auctioneer, and the interaction moves to its final and terminating state, s_4.

Fig. 2. An auction scenario: the interaction model

TC_1: As an agent, I only trust deadlock free interaction protocols.

TC_2: As an auctioneer, I only trust the interaction if it enforces truth telling on the bidders (i.e. bidders cannot be better off if they bid a value other than their true valuation).

TC_3: If from previous experience I know that DVDs from auctioneer A are not original, then A is not trusted in delivering good quality DVDs.

TC_4: If the auctioneer A is not trusted in delivering good quality DVDs, then it is not trusted in delivering CDs.

TC_5: I trust agent A to take the role of the auctioneer only if it has decent ratings and holds a good reputation, while stressing its most recent ratings.

Fig. 3. An auction scenario: the agents' trust constraints (TC)

3 Trust Models for Multiagent Systems

3.1 Available Trust Models: an Overview

Before presenting our contextualised trust model, we start with an overview of available trust models and mechanisms. Ramchurn, Hunyh, and Jennings (2004) divide these models and mechanisms into two categories (Fig. 4). The first deals with trust at the individual level and the second at the system level. As a result, the first lays the burden of computation on the agent and the second on the system.

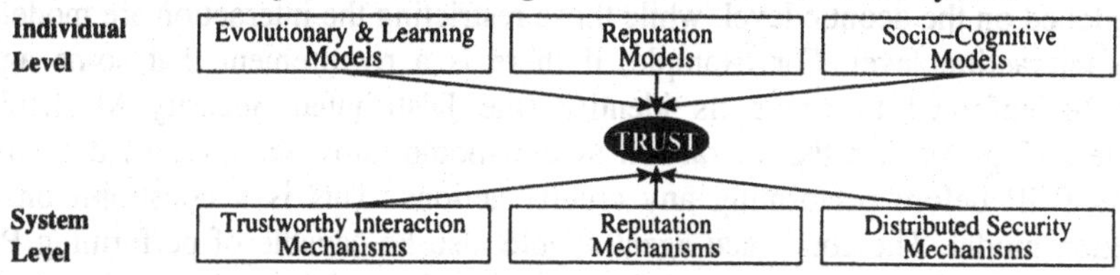

Fig. 4. An overview of MAS's trust models and mechanisms (Ramchurn et al. 2004)

Trust models on the individual level address the agent's trust in others. In evolutionary and learning models, agents interact with others over a period of time, and eventually learn from their past experience. In open systems, however, it is hard for one agent to have enough experience with all others. In this case, the agent may rely on others' experience by seeking the reputation of the agent in question. Other models that do not rely on past experience are the socio-cognitive models, where trusting an agent to perform correctly relies on one's knowledge of the agent's capability, willingness, persistence, or motivation.
System level trust mechanisms focus on driving agents to perform correctly, hence preserving a certain level of trustworthiness in the system. One method of enforcing this is through interactions. For example, while a Vickrey auction encourages truth telling on the bidders' side, English auctions encourage truth telling on the auctioneer's side. Reputation mechanisms may also be enforced by the system encouraging agents, for instance, to rate each other at the end of each interaction. Moreover, for preserving a certain level of trustworthiness, there are also the more traditional distributed security mechanisms which deal with identity proof, access permission, content integrity, content privacy, etc.

3.2 Proposed Trust Model: a Contextualized Model

The trust mechanisms presented above may be favoured for one scenario yet proved to be exhaustive and even useless for another, hence the need for a contextualised trust model. Since agents group themselves at runtime into different interactions, we leave it to the agents to decide whether the interaction it is about to join and the agents it will be collaborating with are trustworthy or not. To do so, our contextualised model focuses on the agents' dynamic trust requirements, as opposed to the system's fixed trust requirements. Ramchurn et al.'s (2004) categorisation splits the different approaches for modelling trust into the following two levels: the individual level and the system level, which represent the two bodies (the agents and the system) responsible for enforcing trust. Our approach proposes a slightly different categorisation with the following two levels: the local deontic level and the global interaction level, which represent the constraints on the two bodies (the agents and the interaction). The constraints here are specified by the agents themselves. In what follows, we provide a comparison between these two trust models.

3.3 Proposed Versus Existing Trust Models: a Comparison

Figure 5 provides examples from Ramchurn et al.'s (2004) trust mode mapped into our proposed model. Notice that trust issues that restrict agents actions are modelled on the deontic level, while those restricting the interaction are modelled on the interaction level. For example, if there is a requirement that some agent should be enforced to prove its identity (the Distributed Security Mechanism example of Fig. 5), then the interaction model should allow the concerned agent to send its PGP before performing any crucial actions. This is a constraint on the interaction model. The concerned agent should also be capable of performing PGP encryptions. This is a constraint on the agent's deontic model. Furthermore, note that while an agent might find this requirement crucial for one scenario, the same trust requirement may become useless for another, hence the need for a contextualised trust model.

4 Trust Specification

The agents' trust constraints are constraints either on the interaction or on the agents. In the latter case, the constraints could either be general constraints on the agent or more specific constraints, such as trusting the agent in performing a specific action. The proposed trust policy language (Fig. 6) is used for specifying such constraints. The following trust constraints provide an example on how to specify some of the constraints of Fig. 3:

$$\begin{aligned}&trust(interaction(I), +) \leftarrow\\&\quad (bid(bidder, V) \rightarrow win(bidder, P_v)) \wedge\\&\quad (bid(bidder, B) \rightarrow win(bidder, P_b)) \wedge B \neq V \wedge\\&\quad P_b \nless P_v.\end{aligned} \qquad (TC_2)$$

$$\begin{aligned}&trust(a(auctioneer, A), +) \leftarrow\\&\quad rating\_count(a(auctioneer, A), Total) \wedge Total > 50 \wedge\\&\quad rating\_average(a(auctioneer, A), Average) \wedge Average > 0.8 \wedge\\&\quad rating\_latest(a(auctioneer, A), 15, Latest) \wedge Latest > 0.95.\end{aligned} \qquad (TC_5)$$

Property TC2 specifies that the interaction (for example, that of Fig. 2) is trusted only if the bidder cannot do any better than bidding its true valuation. This is specified via the temporal property that states: if the bidder bids its true valuation V, then it could win the item for the price Pv, and if it bids B, which is different than its true valuation V, then it could win the item for the price Pb, and it is always the case that Pb is not less than Pv (i.e. the agent cannot do any better than bidding V).

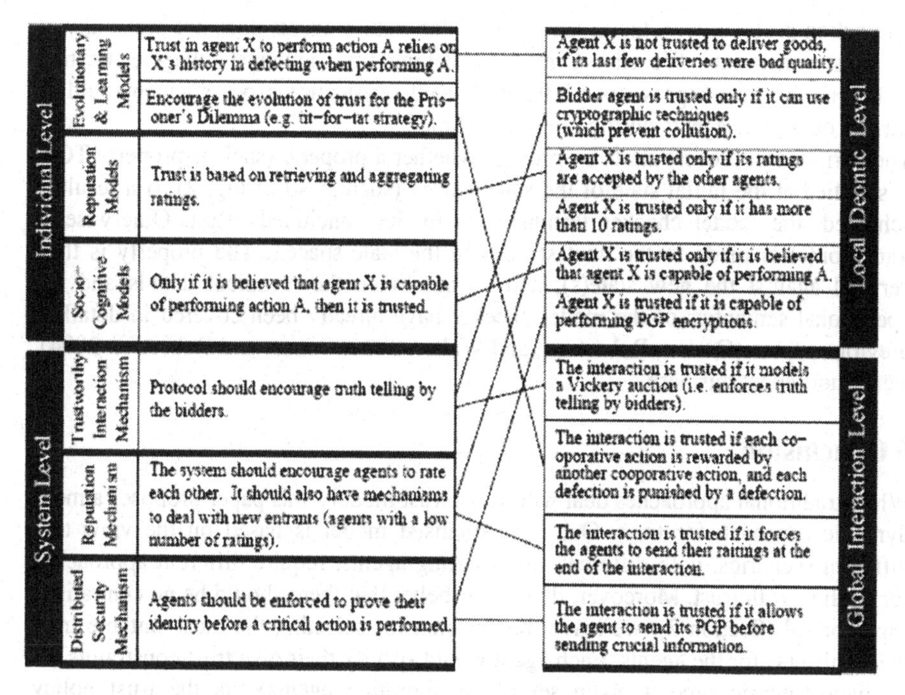

Fig. 5. Mapping between Ramchurn's (2004) categorisation of trust and our proposed model

---

Syntax:

$TrustRule$	::	$TrustSpecs$ [$\leftarrow$ $Condition$]
$TrustSpecs$	::	$trust(interaction(I), Sign)$ | $trust(Agent, Sign)$ | $trust(Agent, Sign, Action)$
$Agent$	::	$a(Role, Id)$
$Sign$	::	$+$ | $-$
$Action$	::	MPA | $N\text{-}MPA$ | $TrustSpecs$
MPA	::	$Message \Rightarrow Agent$ | $Message \Leftarrow Agent$
$Condition, I, Role, N\text{-}MPA, Message$	::	$Term$

Semantics:

Trust rules might be constrained by additional conditions: $TrustSpecs$ [$\leftarrow$ $Condition$]. The interaction's trustworthiness is modelled by $trust(interaction(I), Sign)$, while the agent's trustworthiness is modelled by $trust(Agent, Sign)$, where $+$ and $-$ values of $Sign$ are used to model trust and distrust, respectively. Only if the agent is trustworthy, it can engage in an interaction. Trusting or distrusting agents to perform specific actions is modelled via: $trust(Agent, Sign, Action)$. Actions could either be message passing actions (MPA) — such as sending ($Message \Rightarrow Agent$) or receiving ($Message \Leftarrow Agent$) messages — or non-message passing actions ($N\text{-}MPA$) — such as performing computations. Actions may also take the form of another trust rule (see $TrustSpecs$ in the $Action$ definition). The latter supports the delegation of trust since it permits the specification of whether an agent's trust is to be trusted or not.

Fig. 6. Trust policy language: syntax and semantics

Property TC2 is essentially a constraint on the interaction. On the other hand, property TC5 specifies that the auctioneer A is trusted only if has an acceptable number of ratings (e.g. more than 50) with at least 80% of those being positive, going up to 95% for the latest ratings (e.g. the last 15). Property TC5 is an example of constraints on the agents, rather than the interaction.

5 Trust Verification

After specifying trust constraints, agents should then choose an appropriate interaction protocol. This is achieved by allowing the agents to verify that the given interaction protocol with the given set of collaborating agents does not break any of the trust constraints. We choose model checking from amongst other verification techniques because it provides a fully automated verification process which could be carried out by the agents during interaction time.

The verification process starts by proving whether a property (such as property TC2) is satisfied at the initial state of the state-space[1] (such as s0 of Fig. 2). If a result is achieved, the model checker terminates with the concluded result. Otherwise, a transition(s) is made to the next state(s) in the state-space[2]. The property is then verified against the new state(s), and so on. Since the technical details and the operational semantics of the model checker have already been covered in detail in previous papers (Osman, Robertson, and Walton 2006; Osman and Robertson 2004), we do not repeat this here.

6 Conclusion

While traditional approaches deal with fixed trust models, this paper proposes a more dynamic approach for trust. Our contextualised model is based on the view that different scenarios, with different collaborating agents, require different approaches for dealing with trust. Moreover, due to our belief that there should be no centralised trust control (Section 2), we leave the decision for selecting the most suitable trust mechanism(s) for the agents. Each agent would specify their own trust constraints for a given scenario (and a given set of collaborating agents) via the trust policy language presented in Section 4. We also show that agents are then capable of verifying (via the model checker of Osman et al. 2004) whether their trust constraints will be broken if they engage in the given scenario.

References

Osman, N. and Robertson, D. (2007) Dynamic verification of trust in distributed open systems. In *Proceedings of the 20th International Joint Conference on Artificial Intelligence*.

Osman, N., Robertson, D. and Walton, C. (2006) Dynamic model checking for multiagent systems. In *Proceedings of Declarative Agent Languages and Technologies*.

Ramchurn, S.D., Hunyh, D. and Jennings, N.R. (2004) Trust in multi-agent systems. Knowledge Engineering Review.

Robertson, D. (2004) A lightweight coordination calculus for agent social norms. In *Proceedings of Declarative Agent Languages and Technologies workshop*.

---

[1] *Satisfaction is verified by applying the modal μ-calculus proof rules (Osman et al. 2006).*

[2] *Transitions are based on the LCC (Robertson 2004) transition rules (Osman et al. 2006).*

Evaluation of Security Mechanisms for Virtual Organizations

Jake Wu[1] *and Panos Periorellis*[2]

1 Newcastle University, School of Computing Science, jake.wu@newcastle.ac.uk
2 Newcastle University, School of Computing Science, panayiotis.periorellis@ncl.ac.uk

Abstract. GOLD project is concerned with dynamic formation and management of virtual organizations in order to exploit market opportunities. The project aims to deliver the enabling technology to support the full lifecycle of such VOs. A set of middleware technologies have been designed and implemented to address issues such as trust, security, contract management, monitoring and information management for virtual collaboration between companies. In this paper we will showcase some of the more general requirements for authentication and authorization in GOLD VOs. In conjunction with these requirements we evaluate some of the more popular tools that are currently available in dealing with these issues, together with our own approach in addressing these problems.

1 Introduction

Authentication and authorization mechanisms are integral to the operation of any virtual organization (VO). The GOLD project spent a considerable amount of time and effort gathering a full set of requirements (Periorellis, Townson, and English 2004) that address the needs of VOs in terms of such mechanisms. We have identified a need for flexible, interoperable solutions that are capable of dealing with the cross-organizational nature of virtual organizations, as well as its dynamics in terms of composition. The tools that we discuss in this paper are open source solutions that are popular amongst the e-Science community. We attempt to evaluate these tools in conjunction with a set of requirements for authentication and authorization in VOs that we will discuss in the next section. The paper presents a comparative evaluation of four tools and associated frameworks, highlighting their role in tackling the requirements discussed below.

2 Requirements

Virtual organizations bring together a number of autonomic organizations to assess a market opportunity. The duration of this collaboration can be brief, or require a

longer life-span. It is certain that during the life-span of a VO, the parties involved will be required to share resources therefore; access to those resources may require the crossing of organizational boundaries. In terms of authentication, participants of a VO are expected to have implemented their own security mechanisms to protect resources within their boundaries. A participant may require access to several resources scattered across several organizational boundaries. The requirement here is for multiple logins across a number of sites. The way we solve this problem is by involving trusted third parties that provide security assertions as and when they are requested. A security assertion is token such as a signed credential. The token must allow single sign on to be achieved without dictating to the VO participants who the trusted third parties should be. This approach minimizes the identity flows in the system, thereby providing better protection of the user's personal information. Another possible solution is federation. This involves parties agreeing on pre-determined trust relations between service providers and identity providers. Beyond federation we also need to provide flexible protocols that allow participants to validate security assertions using authorities trusted by these participants.
With regard to authorization, given the dynamic nature of VOs, and the sensitivity of sharing information, static rights assignment is not sufficient to capture all the eventualities. Also VO participants are not expected to hold a set of permissions that last throughout the VO life-span. It is more likely that companies will agree limited access, or gradual access to their resources depending on progress of the collaboration activity. In GOLD we aim to be able to enable VOs to define conceptual boundaries around projects and tasks, so that roles and permissions can be scoped. It must allow for the dynamic activation and deactivation of permissions and roles based on progress monitoring of projects and also the tasks performed, compared with the established contracts. There is, therefore, a need to adopt fine grained access control mechanisms. In GOLD, the simple subject-object permissions model upon which Role Based Access Control (RBAC) is based is not sufficient. Instead, GOLD requires fine grained permissions for specific instances of roles as well as instances of objects. Also, GOLD needs to support the delegation of roles and privileges between different levels of authorities. Given the wide range of policies for access control within a VO and the fact that no single authority governs these policies, GOLD also requires validation. This is aimed at eliminating any logical inconsistencies in policy/permissions prior to any workflow enactment.

3 Tools and Evaluation

In this section we briefly introduce four tools that target similar issues of authentication and authorization. These tools are PERMIS, OASIS, Shibboleth, XACML. We will present these tools and their features individually in this section. In the subsequent section we will provide a detailed evaluation of these tools in conjunction with our requirements.

3.1 PERMIS–PrivilEge and Role Management Infrastructure Standards Validation

PERMIS was funded by Information Society Initiative in Standardization and developed by Salford University. PERMIS dictates a typical role based access

control (RBAC) model and aims to provide a solution for managing user credentials when accessing target resources. It is used in electronic transactions in governments, universities or businesses for solving the "authentication of the personal identity of the parties involved" and the "determination of the roles, status, privileges, or other socio-economic attributes" of the individual, through the use of X.509 attribute certificates (ACs) (PERMIS 2001). ACs are widely used to store users' roles and XML-based authorization policies for access control decisions in PERMIS. They are digitally signed by the issuers and stored in the public repositories. Chadwick & Otenko (2003) have summarized basic features of PERMIS which include being a mechanism for identifying users, specifying policies to govern the actions users can perform, and making access control decisions based on policy checking.
In short, as an RBAC system, PERMIS bases all the access control decisions on the roles for users and policies. Roles and policies are respectively stored in X.509 ACs, which are then protected by digital signature and kept in the public repository. In the PERMIS architecture, a user makes an access request via an application gateway. In the application gateway the Access Control Enforcement Function unit authenticates the user and asks the Access Control Decision Function (ADF) unit if the user is permitted to perform the requested action on the target service provider. The ADF accesses LDAP (Lightweight Directory Access Protocol) directories to retrieve the policy and role ACs, and then makes a granted or denied decision. PERMIS defines its own policy grammar. The typical components that PERMIS XML policy comprises are defined by Bacon et al (2003), Chadwick & Otenko (2003).

3.2 OASIS-An Open, role-based, Access control architecture for Secure Interworking Services

OASIS is developed by the Opera research group in Cambridge Computer Lab (OASIS 2003) and is an RBAC system for open, interworking services in a distributed environment, with services being grouped into domains for the purpose of management. The aim of OASIS is to provide a standard mechanism for users and services in a distributed environment to interwork securely with access control policies enforced. The OASIS system is based on Role Membership Certificates (RMC) issued to the users and Credential Records (CR) stored on the servers. One focus of OASIS is the dynamic role activation. For example, in order for a user to possess a role, there may be a set of role activation conditions that must be satisfied. These conditions may include requirements for prerequisite roles, and any other constraints. Roles can be activated or deactivated dynamically as situations arise. OASIS uses appointment certificates (ACs) for associating privileges persistently with membership of some roles and for handling delegation of rights between users. These certificates can be issued to users by some roles with the particular functions and they can serve as a form of credential to satisfy the role activation conditions for a user to activate one or more other roles. Some other key differences are summarized by Bacon et al (2003) between OASIS and other typical RBAC schemes. A number of RBAC models have been proposed by Sandhu et al (1996).

3.3 Shibboleth –Federated Identity

Shibboleth is a project developed at Internet2/MACE (1996-2005). It is an identity management (user attributes based) system designed to provide federated identity and aims to meet the needs of the higher education and research communities to share secured online services and access restricted digital content. It focuses on providing a way for a user using a web browser to be authorized to access a target site using information held at the user's security domain. It also aims to allow users to access controlled information securely from anywhere without the need of additional authentication process. "Shibboleth is developing architectures, policy structures, practical technologies and an open source implementation to support inter-institutional sharing of web resources subject to access control"(Shibboleth 2005). The design of Shibboleth was based on a few key concepts:

a. Federated Administration.
b. Access Control Based on Attributes.
c. Active Management of Privacy.
d. Standards Based.
e. A Framework for Multiple, Scaleable Trust and Policy Sets (Federations).
f. Has defined a standard set of attributes.

The Shibboleth system includes two main components: Identity Provider (IdP) and Service Provider (SP) (Shibboleth 2005; Morgan, Cantor, Carmody, Hoehn and Klingenstein 2004; Vullings, Buchhorn and Dalziel 2005). IdP is associated with the origin site. SP is associated with the target site. These two components are usually deployed separately but they work together to provide secure access to web based resources. Shibboleth aims to exchange user attributes securely across domains for authorization purposes and PKI is the foundation to build the predetermined trust relation between Shibboleth components, i.e. IdPs and SPs, of the federated members. Vullings et al (2005) notes that the main assumption underlying Shibboleth is that the IdP and the SP trust each other within a federation. Some main features of this federation based system are summarized by Morgan et al (2004). Shibboleth uses OpenSAML APIs to standardize message and assertion formats, and bases protocol bindings on SAML.

3.4 XACML-eXtensible Access Control Mark-up Language

XACML is an XML based Web Service standard for evaluating security sensitive requests against access control policies. It provides standard XML schema for expressing policies, rules based on those policies and conditions. It also specifies a request/response protocol for sending a request and having it approved.

The XACML specification also defines architecture for handling the entire lifecycle of a request, from its creation through to its evaluation and response. Several components such as the Policy Enforcement Point (PEP) and Policy Decision Point (PDP) transform requests into the standard format before they are evaluated using a rule combining algorithms (SUNXACML 2004). The benefits of using XACML as written in the specification are summarized in SUNXACML (2004). XACML allows fine-grained access control and is based on the assumption that a user's request to perform an action on a resource under certain conditions needs an "allow" or "deny" decision.

4 Discussion

The requirements section of this paper described how virtual organization dynamics can affect decisions regarding the implementation of authentication and authorization mechanisms. In this section, we evaluate the tools based on a few elements we draw from the requirements. Table 1 (Evaluation of tools) gives an overview of these requirements and describes how other tools tackle these problems.

4.1 Authentication and SSO

The GOLD system supports SSO via the usage of SAML assertions and related protocols as specified by Liberty Alliance (Liberty Alliance Project 2003), using OpenSAML APIs (OpenSAML 2005). PERMIS offers an RBAC authorization infrastructure. However, PERMIS has been built to be authentication agnostic (Bacon 2003). When a user wishes to access an application controlled by PERMIS, the user must first be authenticated by the application specific AEF. OASIS does not have support for authentication; it is entirely an access control system. Shibboleth implements the SSO mechanism via the use of SAML and the concept of federated identity, therefore users are only required to sign on once at home organizations. In the GOLD authentication system, we support a variety of tokens, e.g. X.509 digital signature/encryption and SAML assertion. In PERMIS, when authenticating a user the AEF could use digital signatures, Kerberos, or username/password pairs (Bacon et al. 2003). Shibboleth mainly uses the SAML standard to construct its tokens through OpenSAML APIs which is also developed at Internet2 (Internet2/MACE 1996-2005) while other tokens are also supported. The standardization that SAML offers largely facilitates the exchange of security information about users in trust domains and fits nicely with the other WS-* standards utilized in the GOLD system.

4.2 Privacy

Privacy is also an essential issue that a security system needs to address. In the GOLD system, a user is issued with a SAML assertion, or a GOLD context id in the prototype, and only this assertion or context id flows between the GOLD participants. A user is only authenticated once with his private information and does not need to provide this information repeatedly. PERMIS does not provide privacy protection. Instead, PERMIS uses public repositories to store the attribute certificates, which compromises the user's privacy (Bacon et al. 2003). In OASIS the ACs are not publicly visible when issued and presented. When ACs go through communication channels they are encrypted under SSL, therefore the privacy is ensured. In Shibboleth, fairly active management of privacy was in place when the system was designed and the users have full controls over what personal information is released and to whom.

4.3 Federation and broker style trust

Federation is an indispensable part of the GOLD architecture and offers the GOLD participants the choice of deciding whether they want to trust the identity providers based on a pre-determined trust relationship. PERMIS and OASIS have been developed in a distributed environment (PERMIS 2001; OASIS 2003). Undoubtedly federation is the paramount issue in the Shibboleth system. However the GOLD system also offers participants a choice of finding alternative independent parties in the case of parties disagreeing on which authorities they trust, without sacrificing any traceability or accountability of credentials. This broker style trust is not currently supported in any of other aforementioned tools.

4.4 Dynamic activation and deactivation of access rights

Central to the authorization requirements is the need for dynamic activation and deactivation of access rights. It was noted earlier that the GOLD framework supports this by enabling VOs to define conceptual boundaries around projects and tasks such that the roles and permissions can be scoped. Ongoing decisions can be made along with the progresses of projects or tasks in relation to the dynamic activation and deactivation of the access rights. A high level of granularity is provided in this context. These are all achievable by taking advantage of the standardization and flexibility that XACML can offer. OASIS (2003) provides the means for

	PERMIS	OASIS	Shibboleth	Our approach
Single-sign-on	Supported (built-in in authorization system)	N/A	Supported	Supported
User privacy	Not Supported	N/A	Supported with active privacy management	Supported (good privacy maintenance and identity control)
Federation	Distributed environment	Distributed environment	Supported	Supported
Broker style trust	N/A	N/A	Not supported	Supported
Variety of tokens	Supported in authentication (digital signature, kerboros, Username/password)	N/A	Supported (mainly use SAML)	Supported (X509, SAML assertion, etc)
Standardised language	N/A	N/A	SAML standards based	SAML standards based and use of XACML
Dynamic activation and deactivation of access rights	Not supported (persistent policy rights with the use of static attribute certificates)	Dynamic role activation with the use of appointment certificates	N/A (attribute-based authorization)	Supported
Policy delegation	Supported in current release; not in previous releases	Not supported (use appointment instead)	N/A	Supported

Table. 1. Evaluation of Tools

dynamic role activation as discussed in the earlier section 3.2. In contrast, the role and policy rights assignment in PERMIS are rather persistent and the attribute certificates PERMIS uses are a static representation. According to the PERMIS developers, in the current release a role is revoked by explicitly deleting the role AC from the LDAP directory using an LDAP browser/admin tool. Shibboleth focuses on attribute-based authorization and does not mention the dynamic use of policy rights.

4.5 Policy delegation

Also in GOLD, we support the delegation of authorities, where the source of authorities can delegate roles/privileges/rights to the subordinate authorities. The X.509 standard, which PERMIS is based on, specifies mechanisms to control the delegation of authority from the source of authority to subordinate attribute authorities. The delegation was not fully supported by the PERMIS implementation in the previous releases. However, it is currently supported by the recently developed release according to the developers. A role holder may delegate his/her role to another individual, without the need to have permission to alter the privileges assigned to that individual. Furthermore PERMIS supports role hierarchies. With role hierarchies privileges of subordinate roles can be inherited by superior roles, and a role holder can delegate just a subordinate role instead of the entire role (Chadwick and Otento 2002). OASIS uses appointment as introduced earlier, as opposed to the privilege delegation.

5 Conclusion

GOLD middleware offers a set of services that can be used to assist in the formation, operation and termination of VOs. The aim of the project and the proposed architecture is to offer VO developers the flexibility to configure the VO according to their requirements without imposing too many constraints or imposing what and how it should be done. This paper looked at VO requirements in terms of authorization and authentication, taking into account the dynamics of a VO, the levels of distrust, the need for federation as well as rights delegation. We have evaluated the four tools available under open source licensing which address similar issues. We give practical assessments on what these tools do and how they address issues that we are concerned in relation to the requirements we elicited from the project. In GOLD we devised a security infrastructure for authorization and authentication which is based primarily on current WS and OASIS (Organization for the Advancement of Structured Information standards) standards. The infrastructure supports privacy of a user's own information as long as there is a traceable link between the federated identity (valid only within the VO) and their real credentials as the federated identity does not identify the real identity of the participant. The adoption of automated policy creation, maintenance and enforcement with standardization offers the flexibility of coping with the dynamic VOs. It is undesirable to impose unnecessary constraints on the autonomous parties by dictating the specifics of the various supporting technologies the parties are required to deploy in order to participate in a VO. The parties should retain their autonomy as to how their resources should be protected. The high expressiveness and flexibility of the XACML policy language facilitates the centralized policy storage and the granularity of policy expression whereby a fine grained access control mechanism is enabled. A current limitation of such system is that the gap between the actual users and policy language expression is not fully resolved. A well designed front-end is essential to allow users to easily interact with the system hence to achieve better usability. This for instance may involve natural language processing. The GOLD provides the core architectural components (storage, security, co-ordination, regulation) as a set of services. A well chosen security mechanism should also facilitate the integration with other components.

References

Bacon J. et al. (2003) Persistent versus Dynamic Role Membership, 17th IFIP WG3 Annual Working Conference on Data and Application Security, No. 17, pp. 344-357

Chadwick D. and Otento A.(2002)RBAC Policies in XML for X.509 Based Privilege Management, in Security in the Information Society: Visions and Perspectives: IFIP TC11 17th Int. Conf. On Information Security, Cairo, Egypt. pp. 39-53

Chadwick D., Otenko S. (2003_1) A comparison of the Akenti and Permis authorization infrastructures, Proceedings of the ITI First International Conference on Information and Communications Technology (ICICT 2003) Cairo University, pp. 5-26.

Internet2/MACE (1996-2005) Internet2-Middleware Architecture Committee for Education, http://middleware.internet2.edu/MACE/

Liberty Alliance Project (2003) Introduction to the Liberty Alliance Identity Architecture.

Morgan R. L., Cantor S., Carmody S., Hoehn W. and Klingenstein K. (2004) Federated Security: The Shibboleth Approach, Educause Quarterly, Vol. 27, No. 4.

OASIS (2003) An Open, Role-based, Access Control Architecture for Secure Interworking Services, Cambridge University EPSRC project, http://www.cl.cam.ac.uk/Research/SRG/opera/projects/

OpenSAML (2005) An Open Source Security Assertion Markup Language implementation (Internet2) http://www.opensaml.org/

Periorellis, P., Townson, C. and English, P. (2004) CS-TR: 854 Structural Concepts for Trust, Contract and Security Management for a Virtual Chemical Engineering, Newcastle Univ.

PERMIS(2001)Privilege and Role Management Infrastructure Standards Validation, http://www.permis.org

Sandhu R., Coyne E., Feinstein H. and Youman C. (1996) Role-Based Access Control Models, IEEE Computer, Volume 29, Number 2.

Shibboleth (2005) Shibboleth Project, Internet2/MACE, http://shibboleth.internet2.edu

SUNXACML (2004) Sun's XACML Implementation, http://sunxacml.sourceforge.net

Vullings E., Buchhorn M. and Dalziel J. (2005) Secure Federated Access to GRID applications using SAML/XACML.

Non-Repudiable and Repudiable Authentications in E-Systems

Song Y Yan and Tim French

Institute for Research in Applicable Computing
The University of Bedfordshire Park Square Luton, Bedfordshire LU1 3JU
fsong.yan@beds.ac.uk; tim.french@beds.ac.ukg

Abstract: Authentication and non-repudiation are intimately related to each other. In fact, a good authentication scheme must have the property of non-repudiation, otherwise the authentication scheme may not be very useful in practice since the signatory can deny his signature later. This may, however, not be the case for some advanced e-voting systems where non-repudiation should in fact be avoided whenever possible, since for the purpose of privacy, the e-voter does not want to disclose his authorship. Nevertheless, the authorship is veriffiable by the author if needed. In this paper, we propose two implementations for two types of authentication: 1) non-repudiable authentication in a scientiffic computing environment: computing the complex zeros of the Riemann ^{3}-function or verifying the Goldbach's conjecture, and 2) repudiable authentications in an e-voting environment. The security of the ffirst implementation is based on the intractability of the Elliptic Curve Discrete Logarithm Problem (ECDLP), whereas the second is based on the intractability of the Quadratic Residuosity Problem (QRP).

Keywords: Non-repudiable and repudiable authentication, intractabil- ity, discrete logarithm problem, elliptic curve discrete logarithm problem, and quadratic residuosity problem.

1 Introduction

Traditionally, authentication and non-repudiation are intimately related to each other:

Authentication ⇔ Non-Repudiation

However, in many cases such as in e-voting systems, we need to adopt a rather different authentication scheme - the repudiable authentication scheme, in which the author/owner of as e-vote can deny his authorship/ownership. In this paper, two implementations for the two types of authentication will be proposed, discussed and analysed:

1. Non-repudiable authentication in a scientific computing environment (computing the complex zeros of the Riemann ^{3}-function),

2. Repudiable authentication in an e-voting environment.

The security of the first implementation is based on the intractability of the Elliptic Curve Discrete Logarithm Problem (ECDLP), whereas the second is based on the intractability of the Quadratic Residuosity Problem (QRP).

2 Non-Repudiable Authentication

In the theory of numbers, there are many difficult unsolved problems that have remained opened for many years; Goldbach's conjecture and Riemann's hypothesis are just two of them. From a computational point of view, the solutions to these problems are mainly finding counter-examples. For example, the famous Goldbach's conjecture states that every even number is the sum of two prime numbers such as 8 = 3 + 5. If anyone is lucky enough to find an even number which is not expressible as a sum of two prime numbers, then the problem is solved. Clearly, to find such a counter-example, a more powerful computing system such as the grid system is needed; the current record is that Goldbach's conjecture is true up to 5 ¢ 1017. Now let us consider Riemann's hypothesis in the grid computing environment, aiming at finding the complex zeros of the Riemann ς function:

$$\varsigma(s) = \sum_{n=1}^{\infty} \frac{1}{n^s}$$

Where $s = \sigma + it$ with $i = \sqrt{-1}$ and $\sigma, t \in \Re$ Riemann conjectured in 1859 that all the complex zeros lie on the vertical line of $\sigma = 1/2$

This is the infamous Riemann's hypothesis, a one-million US dollar prize problem [2]. The first five such zeros above the x-axis are shown in Figure 1 (with the other 5 conjugate complex zeros below the x-axis). Suppose that a grid service provider has a powerful computing system that can locate the zeros efficiently and can provide its system to authorized users for the purpose of locating more zeros. Now suppose further that Alice is one of the authorized users (subject to paying the fees of the expensive CPU time), whereas Bob is not allowed to use the system even if he is willing to pay the fees. A solution to this problem is that each time when Alice requests to use the system, she digitally signs her request, generated by her private key priAlice. Since Alice is the intended user, the system has her public key pubAlice on file, so the authenticity of Alice can be verified and the access to the system can be granted. More importantly, later when a bill is sent to Alice for the fees of the expensive CPU time she has consumed, she cannot later deny paying the fees, since she is the only one who can produce her signature on her request by her private key pri_{Alice}.

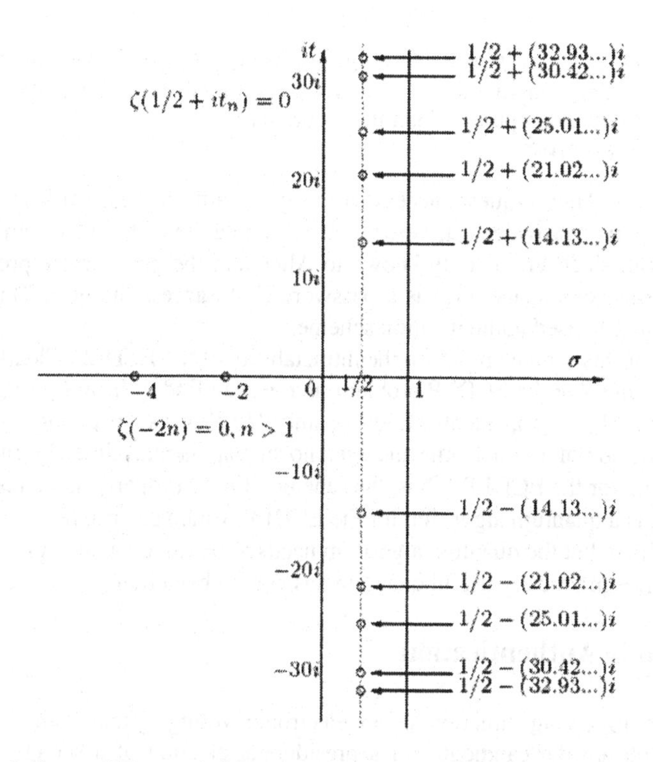

Figure 1. Zeros of Riemann's ³-function (Some Initial Values on the Critical Line)

This is the very necessary condition for her to be granted access to use the system. Of course, Bob cannot access the system, since he is not the intended user and the system of course does not have his public-key pub_{Bob} on file, hence the authenticity of Bob cannot be verified. Subsequently, his request to use the system is not allowed and will be blocked. Related discussion on grid computation of the ³-function can be found in the beautifully prepared conference slides [6] and [7]. Non-repudiable authentication in fact was the most important motivation for the earlier development of modern public-key cryptography by Diffie and Hellman in the 1970s (see [3] and [4]).

To make the authentication of the grid even easier, more efficient and more secure, an elliptic curve analog of the Diffie-Hellman-Merkle key exchange scheme (see [3] and [10]) can be implemented and used in this environment. Let $E : y^2 = x^3 + ax + b$ be an elliptic curve over a finite field F_p, denoted by E=F_p, and $P; Q$ 2 E points on the curve. The ECDLP problem may be defined as follows.

Given ($E; P; Q$), find the integer k such that $Q = kP$: That is, $k = \log_P Q$: We use a variant of the ECDLP, the elliptic curve analog of the DHM key- exchange scheme, called the ECDHM, which may be defined as follows:

Suppose that Alice (A) and the server the grid server provider (S) wish to form a password/ID K_{AS} for Alice to access the grid service. Then:

[1] Both Alice and the grid server provider pre-agree an elliptic curve E and a base point $P \in E$ over a public channel.

[2] Alice computes $k_A = x_A P$ and and sends this to the grid server provider. [3] The grid server provider computes $k_S = x_S P$ and sends it to Alice. [4] Both Alice and the grid service provider compute and get the secure pass-
word/ID: $k_{AS} = x_A x_S P$.

Later, whenever Alice requests access to the grid with the password/ID/certificate k_{AS}, the grid server provider can verify her certificate, and of course will let Alice access the grid, since k_{AS} is only known to Alice and the grid service provider. So Alice is the only one to use k_{AS} as a password/ID to access the grid. This may be regarded as an ID-based authentication scheme.
The security of this scheme relies on the intractability of the ECDLP. Clearly, anyone who can solve the ECDLP problem (that is, can find x_A from $k_A = x_A P$ or can find x_S from $k_S = x_S P$), can also find k_{AS}, since $(E; P; k_A; k_S)$ is public. Unfortunately, no polynomial-time and even no subexponential-time algorithm has been found for the ECDLP. Thus, the scheme, if used properly, is secure. Of course, there is a quantum algorithm for the ECDLP, which can run in polynomial-time, but the quantum algorithm needs to be run on a quantum computer, and a practically useful quantum has not yet been built!

3 Repudiable Authentication

Consider the following situation in an electronic voting system, where both Alice and Bob are the candidates of a presidential election of a society. Now suppose that John wants to vote for Alice but does not want to vote for Bob. For privacy purposes, John does not want to let anybody know that he has voted for Alice regardless of whether or not he actually did vote. This type of authentication is obviously different from the one just discussed in the previous section and is absolutely necessary in an e-voting system. It is called *repudiable authentication*, since one can deny one's own authorship/ownership of a vote. Related work to this type of authentication is discussed in [1]. In this paper, we present a way to implement repudiable authentication based on the Quadratic Residuosity Problem (QSP) [10]. Recall that an integer a is a quadratic residue modulo N if $\gcd(a; N) = 1$ and there exists a solution x to the congruence x2 ≡ a (mod N). The quadratic residuosity problem is defined as follows.

Given a and N such that gcd(a, N) = 1, decide whether or not a is a quadratic residue modulo N. That is, to decide whether or not a ∈ QN.

If N = p is an odd prime, then a is a quadratic residue of p if and only if $a^{(p¡1)=2} \equiv 1 \pmod{p}$, which is easily to be determined. However, if N is an odd composite, then one needs to know the prime factorization of N which is generally known to be intractable, since obviously, a is a quadratic residue of N if and only if it is quadratic residue modulo every prime dividing N. Obviously, if $(\frac{a}{N}) = -1$ then $(\frac{a}{p_i}) = -1$ for some i, and a is a quadratic non-residue modulo N. However, it may well be possible that *a* is a quadratic non-residue modulo

N even if $(\frac{a}{p_i}) = -1$. That is to say:

$$a \in Q_N \Rightarrow (\frac{a}{N}) = 1, \quad a \in Q_N \overset{?}{\Leftarrow} (\frac{a}{N}) = 1, \quad a \in \overline{Q}_N \leftrightarrow\Leftarrow (\frac{a}{N}) = -1$$

This is precisely the intractable case to be considered since to decide the quadratic residuosity, the only method available is to factor N first, which is known to be intractable.

As Martin Hellman, the co-founder of public-key cryptography, mentioned in his landmark article [5]: \a true digital signature must be a number (so it can be sent in electronic form) that is easily recognized by the receiver as validating the particular message received, but which could only have been generated by the sender". In repudiable authentication, we need to add a condition that the digital signature should not only be generated by the sender but also be validated by the sender. Even if the sender signs a document, he can later deny his signature and of course he can verify his signature as well if needed. Our approach to repudiable authentication for an e-voting system may be described as follows:

[1] *Generate the non-repudiable signature* $S = \{s_1, s_2, \dots\}$: The voter, say, Bob, uses his private key to generate his digital signature S for his vote:

Bob Newman	$\Rightarrow$	0215020014052313013	$\Rightarrow$	42527067843532368
$\Uparrow$		$\Uparrow$		$\Uparrow$
Name for Signature		Numerical Form (M)		Digital Signituare (S)

This digital signature was generated and verified by

$$S \equiv M^e \equiv 215020014052313013^7$$
$$\equiv 42527067843532368 \pmod{183297707029260652 47}$$
$$M \equiv S^d \equiv 42527067843532368^{785558871522137263}$$
$$\equiv 215020014052313013 \pmod{18329770702926065247}$$

with

$$ed \equiv 1 \pmod{\Phi(18329770702926065247)}$$

This step is the same as most of the ordinary non-repudiable authentication schemes. If Bob directly sends his vote together with his digital signature to the Election Centre, the centre, of course, can verify Bob's signature by using his public-key which is available at the centre. But Bob does not want to disclose his authorship of the vote, so we need to add some more computational steps in order to produce a repudiable authentication signature.

[2] *Add a repudiable feature to the non-repudiable signature:*

[2-1] Randomly add some extra digits (noises) to the digital signature S so

as to get a corresponding repudiable digital signature $S' = \{s'_1; s'_2;...\}$.

[2-2] Generate a binary string $B = \{b_1; b_2; ...\}$ with each bit corresponding to S', assign 1 to the bit if the corresponding digit appears in both S and S', otherwise, assign 0 to the bit.

[2-3] Generate a string of integers $X = \{x_1; x_2; ...\}$ which is the mixed squares and pseudo squares, using methods in [10] or [9], based on the QRP (or even the kth power residuosity problem (kPRP)). To obtain the X string, we choose $N = pq$ with $p; q$ prime. Find a pseudo random square $y \in \mathbb{Z}_N$ such that $y \in Q_N$ and $(\frac{y}{N}) = 1$. $(N; y)$ can be made public, but $p; q$ must be kept secret. Choose at random a number r_i and compute

$$x_i \equiv \begin{cases} r_i^2 \bmod N & if \quad m_i = 0 \quad \text{(random square)} \\ yr_i^2 \bmod N & if \quad m_i = 1 \quad \text{(pseudo random square)} \end{cases}$$

[2-4] Send $\{(S'; X); E\}$ to the Election Centre, where $(S'; X)$ is the repudiable digital signature and E the e-vote. The author of the e-vote can later deny his authorship of the vote, since S' comprises a random string of digits which is different from S, and which can only be detected and verifying by the author of vote, not anyone else.

[2-5] To verify the signature needed, the author of the vote who knows the trap-door information (the prime factors p and q of the composite modulus N) can show anyone that he is the author of the vote by converting X string back to B string and then remove the noise from S' according to B.

To get back the string $B = \{b_1; b_2;...\}$ from the string $X = \{x_1; x_2; ...\}$, the owner of the vote performs the following operations:

$$b_i \equiv \begin{cases} 0, \text{ if } e_i^p = e_i^q = 1 \\ 1, \text{ otherwise} \end{cases} \quad \text{where} \quad \begin{cases} e_i^p = \left(\frac{x_i}{p}\right), \\ e_i^q = \left(\frac{x_i}{q}\right). \end{cases}$$

[2-6] Remove the noise from S' according to B to get back to S, the digital signature of the owner.

Figure 2 presents just a small example as proof of concept. We shall only show how to find the values of x_3 and x_4 in the above table, all the rest are performed in exactly the same way. We first choose a random pseudo square $y = 1234567$ 2 Z_N such that y 2 Q_N and $(^y{}_N) = -1$. $(N; y)$ can be made public, but $p; q$ must be kept secret. Then choose at random the number $r_3 = 8194920765$ and $r_4 = 17402983$, and compute (note that $b_3 = 0; b_4 = 1$):

$$x_3 \equiv r^2_3 \equiv 8194920765^2$$
$$\equiv 116978103928983633 \pmod{18329707029260652}$$

$$x_4 \equiv yr^2_4 \equiv 1234567 . 17402983^2$$
$$\equiv 181262163650551072 \pmod{18329707029260652}$$

To get back to b_3 and b_4, we perform:

$$e_3^p = \left(\frac{x_3}{p}\right) = 1,\ e_3^q = \left(\frac{x_3}{q}\right) = 1,\ e_4^p = \left(\frac{x_4}{p}\right) = -1,\ e_4^q = \left(\frac{x_4}{q}\right) = -1.$$

Thus, we have:

$b_3 = 0$ since $e^p{}_3 = e^q{}_3 = 1$ and $b_4 = 1$ since $e^p{}_4 = e^q{}_4 = -1$:

Since $b_3 = 0$ and its corresponding digit 1 in S'_3 is noise, it should be removed from S'. However, as $b_4 = 1$ and its corresponding digit 4 should be remain in S'. Clearly, after removing all the noise from S', S' will eventually become S, the true digital signature!

S	4 2 5 2 706 78435 3236 8
↕	↕ ↕ ↕ ↕ ↕↕↕↕↕↕↕↕↕↕↕↕↕↕↕↕↕↕↕↕↕↕↕↕↕↕↕↕ ↕ ↕ ↕
S'	7 9 1 4 8521453228917063784359132323 6 6 8
↕	↕ ↕ ↕ ↕ ↕↕↕↕↕↕↕↕↕↕↕↕↕↕↕↕↕↕↕↕↕↕↕↕↕↕↕↕ ↕ ↕ ↕
B	0 0 0 1 0010010010001110111110000111 1 0 1
↕	↕ ↕ ↕ ↕ ↕↕↕↕↕↕↕↕↕↕↕↕↕↕↕↕↕↕↕↕↕↕↕↕↕↕↕↕ ↕ ↕ ↕
X	x_1 x_2 x_3 x_4 . . . . x_{32} x_{33} x_{34}
↕	↕ ↕ ↕ ↕ ↕↕↕↕↕↕↕↕↕↕↕↕↕↕↕↕↕↕↕↕↕↕↕↕↕↕↕↕ ↕ ↕ ↕
B	0 0 0 1 0010010010001110111110000111 1 0 1
↕	↕ ↕ ↕ ↕ ↕↕↕↕↕↕↕↕↕↕↕↕↕↕↕↕↕↕↕↕↕↕↕↕↕↕↕↕ ↕ ↕ ↕
S	4 2 5 2 706 78435 3236 8

Figure 2. An Example of the Repudiable Authentication

Clearly, anyone who can solve the QRP (or the *k*PRP) can distinguish the pseudo squares from the squares (or the pseudo *k*th powers from the *k*th powers), and hence can verify the digital signature and the authorship of the e-vote. But as everybody knows, solving QRP/kPRP is intractable, thus the author can later deny his authorship of an e-vote, regardless of whether or not he actually did actually vote.

4 Conclusion

In this paper, two types of authentication are proposed and discussed: one non-repudiable, the other repudiable. Both schemes are useful in e-trust systems and applications, with the first being more favourable for commercial service access application and the second for e-voting. The first method is now standard in e-trust, but our implementation seems to be novel and simple. The second opens some new directions in advanced e-trust systems particularly in the e-voting environment.
The security of the non-repudiable authentication scheme is based on the intractability of the DLP and the ECDLP, whereas the security of the repudiable

authentications is based on the intractability of the QSP and/or the kPSP, depending on the implementation. At present, all the problems are computationally hard and no polynomial-time algorithm has been found for any of these problems. Thus, our methods are secure against any known attacks, provided that they are implemented and used properly.

Acknowledgements: The author would like to thank Dr. Adam Vile for his encouragement during the writing of the paper.

References

Y. Aumann and M. O. Rabin, Authentication, Enhanced Security and Error Correcting Codes, *Crypto'98, LNCS 1462*, Springer-Verlag, 1998, pp 299{303.

E. Bombieri, Problems of the Millennium: The Riemann Hypothesis, Clay Mathematics Institute, Boston, 2001.

W. Diffie and E. Hellman, \New Directions in Cryptography", *IEEE Transactions on Information Theory*, **22**, 5(1976), 644{654.

W. Diffie and M. E. Hellman, \Privacy and Authentication: An Introductionp to Cryptography", *Proceedings of the IEEE*, **67**, 3(1979), 393{427.

M. E. Hellman, \An Overview of Public Key Cryptography", *IEEE Communications Magazine*, Landmark 1o Articules, 50th Anniversary Commemorative Issue, May 2002, 42{49.

S. Wedeniwski, ZetaGrid - Computational verification of the Riemann Hypothesis, *Conference in Number Theory in Honour of Professor H.C. Williams*, Banff, Alberta, Canada, May 2003.

S. Wedeniwski, ZetaGrid - Experiences with the Grid for everybody, *Grid Comput- ing News*, Ehningen, Germany, November 2003.

V. Welch, and F. Siebenlist, et al., Security for Grid Services, Proceedings of the 12th IEEE International Symposium on High Performance Distributed Computing (HPDC'03), IEEE Computer Society, 2003, 48-57.

S. Y. Yan, A New Cryptographic Scheme based on the kth Power Residuosity Problem, 15th British Colloquium for Theoretical Computer Science (BCTCS15) , Keele University, 14-16 April 1999.

S. Y. Yan, *Number Theory for Computing*, 2nd Edition, Springer-Verlag, 2002.

Semiotic Models of Trust and Usability for Agent-Managed Grid Services

Tim French[1], Wei Huang[1], Richard Hill[2], and Simon Polovina[2]

1 Department of Computing & Information Systems, University of Luton, Luton, Bedfordshire, United Kingdom
{tim.french, wei.huang}@luton.ac.uk
2 Web & Multi-Agents Research Group, Faculty of Arts, Computing, Engineering & Sciences, Sheffield Hallam University, Sheffield, United Kingdom
{r.hill, s.polovina}@shu.ac.uk

Abstract. This paper seeks to build upon existing work concerning the role of 'soft' issues including trust and usability in the context of Grid services. Previous research has suggested that there is a trust 'gap'. Without seeking to engender intangible trust, usability needs cannot be met effectively. Trust formation is a precursor of usability not only at the point of first contact with Grid services but also as an integral part of the user experience. Whilst tangible security aspects of Grid services are relatively well understood, intangible aspects of trust, are less well developed. The contribution offered here aims to fill the trust 'gap' using Stamper's semiotic ladder. Consequently, a tailored HCI is identified to meet the usability and trust needs of a potentially diverse Grid user community. The use of this approach is accordingly offered as means of satisfying the demands of a user-centric agent to fulfil this role.

1 Introduction

Computational models of trust mechanisms based on the notion of trust have been emerging from the research literature. One reason for this is that traditional security mechanisms are being increasingly challenged by open, large scale and decentralized environments. The Grid is specifically characterized by ad hoc collaborations between geographically distributed resources. An explicit trust management component should ideally aim to go well beyond the 'hard' tangible security aspects of Grid services. Rather, at a higher level of organizational abstraction, wider trust dimensions inherent in Grid partnerships such as partner reputation and organizational culture, quality of service issues such as reliability, provenance, values and ethical concerns, all need to be examined and modeled.
Collaborations within a virtual organization (VO) are facilitated by technological mediators, autonomic agents and entities with minimal need for human intervention.

Grid standards and models to support Grid partner trusted work- flow formation within a decentralized security and risk policy environment have yet to emerge, though several EU funded projects are currently examining relevant trust issues at a VO level of abstraction [3], [11].

However, it is recognized that despite these various initiatives, there is a gap in our understanding concerning mainly 'soft' trust aspects of Grid services and various research initiatives are now underway with the aim of ensuring that emergent Grid computing paradigms are based on a fully articulated set of trust and security protocols and standards. Initiatives such as the SECURE project [9] concentrate on enabling 'trust agents' or 'trust entities' with the power to exchange and verify electronic credentials, gathering local evidence, establishing local access rights to data and shared computing resources.

2 Trust Security and the Grid: Some Current Initiatives

At present, Grid security issues are being addressed mainly from the perspective of existing and established E-commerce security mediators and methods such as; SSL (Secure Sockets Layer) certificates for Grid Security Infrastructure (GSI) user authentication, a set of middleware technologies that are designed to ensure that Grid services are invoked in a secure manner. Most of the attention has thus far been placed upon fine-grained (service level) security issues.

3 Trust and the Grid: What is Missing?

Human trust formation is a far more elusive and subtle process than implied by initiatives or trust architectures such as WS-Trust. In particular, human trust formation involves wider trust contexts: organizational, social, and human cultural factors pre-determine human trust formation and expectations and beliefs [6].

Furthermore, human trust is ultimately an irrational cognitive construct and has a strong emotional component. For this reason we propose that the current models and approaches to autonomic trust formation within the emergent Grid computing paradigm should seek to endow agents and entities with wider, more subtle 'soft' contexts, hence seeking to more closely mimic human to human trust formation.

Previous work within B2C E-Service contexts has established that by deconstructing an interface into its constituent signs and individual design elements, and cross-checking these design elements with consumer trust responses, it is possible to explicitly build and design trust into a user's normative E-Service experience [2], [5].

However, the Grid does present additional problems and complexity in comparison with B2C consumer trust studies of the type exemplified by Egger [2] in which an interface and trust model is tailored to a specific single target group of consumers.

However, much of what has been established so far in B2C E-Service contexts can in principle be re-used. For example, it is known and now generally accepted that the following generic interface design features positively engender intangible trust [2], [5]:

–Prominent trust signs displayed (Brand signs, trust seals, third part accreditation seals);

–Minimal navigation paths engender trust, since users relate goal completion to trust;

–Aesthetics and rendering engenders trust if it clearly matches a user's cultural needs.

With respect to ethical issues, it has been previously established [6], [2], that consumer trust formation in B2C contexts at least, is influenced by the prior reputation of the Service provider. When designing the HCI layer it will be essential for the rendering agent to refer to a model that reflects, perhaps in a crude form, an organizations' reputation, including aspects of their ethical policies. It has previously been established that messages of solidarity, hierarchy and ethics are transmitted to consumers via the HCI component of B2C e- commerce sites [12].

4 Related Work: Grid Semiotics and Trust

Elsewhere, [4] we have proposed that a semiotic trust agent or other advanced social and trust enabled agent is ideally needed that can not only mediate security tangibles but also verify the wider trust domains of VO involved in Grid partnerships. The sequence of activities (below) of such an agent might comprise the following simplified set of meta-level activities:

1. Examine local trust contexts and identify trust principals.
2. Examine local contexts for evidence of trust.
3. Exchange & Check credentials with trust principals and their proxies.
4. Assign or modify a Grid Service trust and VO reputation score (using a suitable algorithm).
5. Reference local security rights and privileges.
6. Reference wider trust contexts and domains.
7. Establish organisational reputation rating matches minimum norms.
8. Enable Grid service.

Clearly, the agent will act as a kind of trust mediator just prior to and at the point of the triggering of a Grid service. Once this point has been reached the Grid semiotic rendering agent will 'take-over' the task of any specific HCI component rendering and adaptation. There is scope for collaboration between these two agents with respect to dynamic information sharing activities. At Grid service termination, control will be passed back to the trusted semiotic agent to terminate the service.

5 Towards a Novel Semiotic Model

Based on Liu's more general approach to 'soft' issues of the Grid [7], [8], this work maps these concerns to Stamper's semiotic ladder [10] so as to instantiate a new variant. Essentially, the novel semiotic trust ladder offers a way of conceptualizing and modeling trust, at a variety of levels of abstraction, by identifying actors, signs and articulating ways in which these mediate all acts of communication. In Table 1, for each layer of the trust ladder, some exemplar trust issues are identified and aligned to the Grid Service lifecycle. By extending this approach it is hoped to develop a fully comprehensive account of trust issues across the entire Grid Service lifecycle, according to certain categories of user.

6 A Trust-Based Usability Model for Grid Services

A generic n-tier architecture can be adapted so as to support a user-centric rendering agent that can use a set of rules, to adapt the HCI component to the needs of a particular culture group [4] or user profile [1]. Figure 1 illustrates how the extended and simplified n-tier model relates to existing Grid service layer(s).

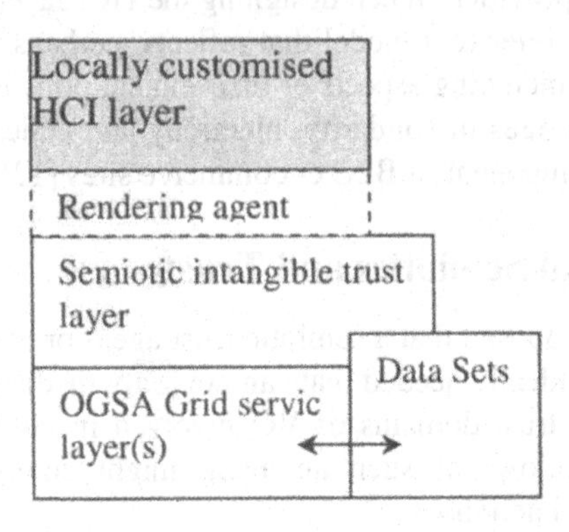

Fig. 1. Extended n-tier architecture

The central notion is to create an additional semantic reference layer encoded in a platform neutral and agent-comprehendable form.

7 Conclusion

The approach to ensuring secure and trustworthy Grid services has previously emphasized hard security aspects, such as WS-Trust, within the OGSA model. This paper has suggested that to achieve clarity of vision the semiotic paradigm may well prove useful in proving a unified conceptual model within which to conceptualize about trust issues in the context of Grid services.
Furthermore, by constructing a trust based usability model for Grid services, based on an extended n-tier architecture, the trust 'gap' can be potentially filled by di erentially rendering of the HCI component so as to match user characteristics and trust and usability needs. By using one unified (semiotic) paradigm to describe VO level trust as well as (in the future) Grid Service level trust, trust issues will be considerably better clarified and articulated that at present.

Exemplar Grid service trust issues	Semiotic trust ladder	Applicability (VO Grid Lifecycle)	Signs for a member of public?	Signs for a scientist?	Signs for a manager?
To what extent does the Service conform to desired VO cultural/cross-cultural norms/legal safeguards?	**Social world trust beliefs and expectations**	Planning stage, run time; User experience layer: Cultural/Social trust/Policy signs	Public branding/is VO branding synergistic with high street/public domain image?	Scientific funding body support/peer group support signs? ESPRC, EU, Royal Society ...?	Legal framework (guarantees, policies, agreements informal and formal?)
Reputation of Grid service provider/ consumer? Any ethical conflicts?	**Pragmatics:** goals, intentions, trusted negotiations, trusted communications	Planning, build, run time User experience layer: Reputation signs	Ethical and Corporate Governance signs	Scientific governance/ Scientific VO reputability signs present?	Signs that legal frameworks are instantiated into machine readable form (can be verified at run time) e.g. legal XML
How reliable, valid are the services and will they meet quality norms?	**Semantics:** meanings, truth/ falsehood, validity	Build and run time User experience layer: Authentication/ validity signs	Any Verisign secure SSL seal program signs	Service guarantees and fault-tolerance?	Legal redress and VO obligation signs in the context of Service quality?
Secure agents: how trusted are they?	**Syntactics:** formalisms, trusted access to data, files, software	Build and run time	Site explanatory narrative written at user level? (explanation of jargon)	Signs of data provenance? Signs of local verification adequate?	Are the tangible mediator layers/signs transparent and verifiable?
Intrusion detection /prevention adequate?	**Empirics:** entropy, channel capacity	Run time	Signs of delays or bottlenecks in accessing system? (user latency)	Signs of delay or unacceptable performance issues?	Messaging/ traffic signs workflow management

Table 1. Macro-Dimensions of a Virtual Organisation and the Semiotic Trust Ladder

However to realise the above, it is necessary to develop a rigorous underpinning that supports the semiotic approach across the myriad domains.
This architecture is intended to more fully and richly simulate human trust formation at the Grid Service level of granularity whilst also referencing wider trust domains, not simply checking and verifying local tangible security credentials.

References

Chu, W., Chen, J., Lee, C., Yang, H (2001), 'Implementing an Agent System Using N-tier Pattern-Based Framework', Procs. 25th International Computer Software and Application Conference (COMPSAC 2001), Invigorating Software Development, 8- 12 October 2001. Chicago, IL, USA. IEEE Computer Society 2001. ISBN 07951127.

Egger, F., (2003) 'From Interactions to Transactions: designing the Trust Experience for Business-to-Consumer Electronic Commerce', Proefshrift, PhD Thesis, Technical University Eindhoven. ISBN 90-386-1778-X.

'Trusted Grid Workflows: Assigning Reputation to Grid Service Providers', ERCIM News, Special Theme: 'Grids: the Next Generation', No. 59, October 2004, 43-44.

French, T., Polovina, S., (2003), 'xCulture for the E-Enterprise: An elementary 4- tier architecture for site localisation', Procs. 2nd BCS HCI & Culture Workshop, Culture and HCI:Bridging Cultural and Digital Divides, Greenwich University, 89- 96. ISBN 1-86166-1916.

Karvonen K., Parkinnen, J., (2001), 'Signs of Trust', Procs. 9th International Conference on HCI, August 5-10, 2001, New Orleans, LA, USA.

Komiak, S., Benbasat, I., (2004), 'Understanding customer trust in Agent-mediated Electronic Commerce, Web-Mediated Electronic Commerce, and Traditional Commerce', Information Technology and Management 5,181-207.

Liu, K., (2003), 'Incorporating Human Aspects into Grid Computing for Collaborative Work', Keynote at the ACM International Workshop on Grid Computing and eScience, 21 June 2003, San Francisco, USA.

Liu, K (2003), 'Semiotics and Grid Computing for Collaborative Work', Seminar for the Software Institute, Chinese Academy of Sciences, Beijing, 8th September, 2003. Available from:http://iel.iscas.ac.cn/english resion/software- institute.ppt256,1, Semiotics and Grid Computing for Collaborative Work.

Di Marzo Serugendo, G., (2004), 'Trust as an Interaction Mechanism for Self- Organising Systems'. Proceedings of the International Conference on Complex Systems (ICCS '04), 2004

Stamper, R., (1976) 'Information in Business Administrative Systems', John Wiley & Sons, UK.

TrustCOM, 6th Framework, disseminated results available from: http://www.eu-trustcom.com/index.php?page=Dissemination

Vile, A., Polovina, S., French, T. (2001) 'eFinance: Architecture, Strategy and Semiotic', Procs. 5th BIT International Conference, Manchester Metropolitan University, UK. ISBN 0-905304-33-0.

Using Trusted Computing in Commercial Grids

Po-Wah Yau[1] *and Allan Tomlinson*[1]

1 Royal Holloway University of London, Information Security Group
{p.yau, allan.tomlinson}@rhul.ac.uk

Abstract. Trust flows in two directions in a Grid environment. The first is from the Grid user to the Grid resource, that is, the Grid trusts that the user will protect confidential information. The second is from the resource to the user, that is, the Grid will protect the user's Grid job and associated data. This paper comments on how Trusted Computing technology can be used to establish trust in both directions, in three types of Grids that may be interest of to commercial organisations.

1 Introduction

Grid computing is a model of distributed computing to enable the pooling of heterogenous resources, for example, CPU cycles, application software, data and its storage (Foster and Kesselman 2004). These resources are accessed by invoking a Grid service that is defined by the resource host. In general, a two-layer administrative domain hierarchy exists. The top layer is an umbrella administrative domain called a Virtual Organisation (VO). The bottom layer consists of separate autonomous domains that host Grid-enabled resources. This separation of domains can be at many levels, for example, groups in a department, departments in an organisation, organisations in a collaborative project etc. Authorised members of the VO will be able to access VO member resources, even though they reside in different domains.

Grid computing is relatively mature in academia (Doyle 2005; Foster and Kesselman 2004) and is attracting interest from industry. However, there are open security issues that are inhibiting the widespread adoption of Grid computing. Specific applications of Trusted Computing (TC), a recent and industry-backed technology (Mitchell 2005), have been proposed to solve certain Grid security problems (Mao, Martin, Jin, and Zhang 2006). In this paper, we concentrate on trust establishment and how to achieve high levels of assurance that may be required in a commercial setting.

This paper is organised as follows. In Section 2 we briefly describe the TC mechanisms that will be useful in a Grid environment. Section 3 describes how current Grid architectures incorporate trust. The rest of this paper discusses trust establishment and the use of TC in three classes of Grid networks: enterprise Grid in Section 4, fixed VO Grid in Section 5, and dynamic VO Grid in Section 6. Section 7 contains some conclusions for the paper.

2 Trusted Computing

Trusted Computing (TC) is an industry-led initiative to provide a variety of security primitives that make use of a hardware root-of-trust called the Trusted Platform Module (TPM) (Trusted Computing Group 2006), which is a tamper-resistant module which is secured bound to the host platform. The features of TC include protected storage within the TPM (mainly for cryptographic keys), remote platform attestation and data sealing; it is the latter two that we are interested in.

The TPM contains a set of Platform Configuration Registers (PCRs) to securely store an audit log of the host platform's boot process. The log consists of a series of platform 'integrity measurements', essentially the hash digest of some software component on the platform. A trusted platform with a TPM can undergo an authenticated boot process – at each stage of the boot process a measurement is taken of the components required for the subsequent stage, before control is passed to those measured components (Grawrock 2006). The platform can attest to its configuration by presenting the platform measurements, digitally signed by the TPM, to a requesting principal. The 'trust' in TC is the assertion that a platform is in a specific configuration. Whether this configuration is one that is 'trustworthy' for Grid computing is an issue we address later. To avoid confusion, we will state that a platform has the 'required configuration' instead of 'trusted' as used in TC terminology.

Several issues with TC platform attestations have been highlighted, in particular, the management of integrity measurements for the seemingly infinite number of configurations that can occur because of software updates and security patches (Cooper and Martin 2006). This is an issue that we will be commenting on later.

A related technology for creating trusted platforms is virtualisation – providing the ability to run multiple 'virtual machines' on one physical platform. This offers several security properties, e.g. process isolation, which is seen as complementary to TC. An example of the union of the two technologies is Intel's LaGrande virtualisation technology (Grawrock 2006), providing hardware support to create secure (and measurable) compartments for virtual machines to operate in.

3 Grid Computing and Trust

The Grid Security Infrastructure (GSI) (Foster and Kesselman 2004) is the de facto architecture that has been adopted by many Grid implementations. Trust is built upon the asymmetric cryptography based SSL/TLS mutual authentication protocols (Dierks and Allen 1999). These rely on a public-key infrastructure (PKI) of which the trust relationships are well understood (Thompson and Olson et al. 2002) – trust is placed in Certification Authorities (CA), their user registration procedures and their ability to protect their private signature keys. The GSI also includes a delegation capability to extend trust (see Section 6). Following the ethos of Grid computing, other authentication protocols can be used, and the WS-Trust specification (Oasis Standards 2007) contains details of specifying how different

credentials can be 'trusted'. The common denominator is that trust is built upon successful entity authentication, i.e. corroborating the identity used.
The issues with relying on identity alone to establish trust in a Grid environment are well documented (Cooper and Martin 2006). Alternative proposals include using attribute certificates (Chadwick, Otenko and Ball 2003) that focus on specifying attributes on which to base access control decisions. We will later see how Trusted Computing can provide another alternative means of establishing trust.
In the hosting environment of the resource/service provider, an authenticated user identity will be mapped to a local identity and user account, and given limited privileges according to either VO or local policy (or both). The local account is then subject to traditional security protections that protect the host environment from malicious processes. In this case, the service provider 'trusts' the user, but only up to a certain degree. Conversely, the user has no option but to completely trust the service provider with any data, software, scripts sent, and the resulting output. This trust asymmetry may not be acceptable within a commercial setting, and is a concern that has been previously expressed (Cooper and Martin 2006; Mao, Martin et al. 2006).

4 Enterprise Grids

Early commercial adoption of Grid has been using enterprise Grid technology, where the VO consists solely of Grid nodes that reside within a single enterprise (Enterprise Grid Alliance 2004), i.e. a single-domain Grid. Such nodes are likely to have strong trust relationships enforced by internal security policies and services, so the trust asymmetry issue highlighted in Section 3 becomes less relevant. Instead, focus is on the Grid access device (user interface) that is used to obtain commercially sensitive information from the enterprise Grid. Information from an enterprise Grid will have potentially high intellectual property value and require additional protection, especially from insider attacks. Private data/results can be 'sealed' using TC, so that only devices with the required configuration can have access. This is of particular importance if access to the single-enterprise Grid is given to external principals, e.g. third-party suppliers in a coordinated development project, or to remote user devices with limited physical security.
Deciding upon the required configuration in an enterprise Grid is a manageable problem. The enterprise is likely to manage software updates for enterprise devices. High risk devices, where users have the autonomy to perform software updates or installations, would probably not be considered for enterprise Grid access.
However, the management of third-party access devices would be outside the enterprise domain. The relationship between the enterprise and third-parties is likely to be enforced with commercial contracts and non-disclosure agreements. One of the conditions of contract could be that access to the enterprise Grid is only possible using TPM-enabled devices with digital rights management (DRM) software installed. The protocols proposed by (Gallery and Tomlinson 2005) and the Trusted Network Connect (TNC) architecture (Trusted Computing Group 2005) provide a means to measure that this configuration is met before allowing a network connection and access to the enterprise Grid.

5 Fixed VO Grids

In fixed VO Grids, Grid nodes reside in distinct administrative domains, have predetermined trust relationships and the VO membership is relatively static.
Such a Grid can be used for multiple enterprise collaboration, each enterprise contributing, through their single-enterprise Grids, propriety, and often specialist, hardware and software (Foster and Kesselman 2004). In a fixed VO Grid, each enterprise Grid represents one node.
In this type of Grid, the protection of enterprise data has to extend outside the enterprise. Jobs and service requests are likely to be submitted directly to the service provider in question, i.e. the node which is hosting the service. Using TC mechanisms, the required configuration of service provider's Grid gateway can be determined before submitting a job to it. If the gateway receives an incoming job that satisfies policy rules and the submitting user has been authenticated etc., then the service provider will accept the job.
However, the gateway will probably not be the platform that the job will be executed on. Consider a Grid service offering computational time. The gateway will pass the authenticated job to a local job manager (e.g. Condor or Portable Batch System), which could also reside on the gateway host. The local job manager will at some point choose an internal worker node to perform the job. This leads to two main issues with securing enterprise data in the host environment.
Firstly, the owner can 'seal' the job and relevant data, so access is only possible if the internal worker nodes meets the required configuration. While it is possible for a worker node to match the requirements of some job owners in order to 'unseal' the job for execution, it would be extremely difficult to do so for all potential VO members.
Secondly, the worker node could be compromised as the job is being executed. A Grid node could be a member of several VOs, potentially running jobs/providing services concurrently for rival organisations. Jobs running on the same worker node would have access to each other's memory space – a vulnerability that could be exploited. While the gateway can be measured to ensure that it is in a state to ensure that jobs are submitted to different worker nodes, local users using the enterprise Grid could potentially introduce or exploit vulnerabilities that could comprise the job (Cooper and Martin 2006).
As discussed in Section 2, platform virtualisation can be used to create protected environments. This would address the second issue highlighted above, by segregating jobs on the same worker node. Virtualisation and TC can also be used to address the first issue of determining required configuration. A user could use a Grid service to download a known secure virtual machine image, and 'seal' data to that image. The virtual machine could include DRM software to provide enhanced protection. The user can delegate rights to the service provider to use the same Grid service to download the same virtual machine image, in order to 'unseal' the user's job for execution. This would also require key migration, from the user's TPM to the worker node's TPM, using a system such as Daonity (Mao, Yan and Chen 2006). Delegation is an issue that we will discuss in the next section.

6 Dynamic VO Grids

There are two issues with fixed VO grids which become even more complicated in dynamic VO Grids: delegation and the architecture needed to support authentication across domains that use different authentication protocols. The vision is that VOs dynamically form to meet the services requested by the Grid user. For example, an enterprise Grid could use a Grid resource broker service to dynamically outsource to other appropriate Grid service providers when additional resources are needed (e.g. for data federation, extra CPU cycles and storage). In order to manage this outsourcing, credentials are delegated using a proxy certificate (Tuecke et al. 2004). This is an X.509 public-key certificate that is generated by a user to extend a trust chain to the entity that the user is delegating rights to. That entity can then make service requests on behalf of the user.

This delegation model could be cumbersome for a dynamic VO Grid as a trust relationship between the user and a service provider may not exist. The service provider may have to 'pull' this relationship using either identity federation or credential translation services, and then make a policy decision to accept the service request or not. Identity federation introduces its own problems, e.g. the need to protect user personal identifiable information. Establishing trust in this way may not provide enough assurances in a commercial setting (see Section 3).

TC can be used to make the delegation model more efficient, by allowing a node to 'push' requirements to delegating Grid services. These requirements could be made in the form of dynamically negotiated Service Level Agreements (SLAs) (Foster and Kesselman 2004). For example, a Grid service may mandate that a requesting entity installs DRM software as part of the SLAs it negotiates. Therefore, the relatively heavyweight mechanisms required for identity federation are not needed because the Grid service has assurance that the data it provides will not be forwarded.

7 Conclusions

A symmetric trust relationship should be established when using Grid computing. In one direction, a Grid service must be able to establish trust in the user and the access device being used. There are many issues with the current methods of establishing trust based on identity. In the opposite direction, the user must be able to trust that the Grid service will not compromise the user's job data. Trusted Computing provides a set of security primitives to achieve this.

Sensitive commercial data can be sealed to a certain platform configuration. Trust is then built on an entity's ability to attest that its platform is in the required configuration to unseal data. In an enterprise Grid, this means determining that an access device, which could have Digital Rights Management software installed, can attest to a state that indicates that it has not been compromised. In fixed VO Grids, a Grid user can seal data on a virtual machine, and delegate rights to a Grid service to retrieve the same virtual machine image (from another Grid service) to unseal data. The delegation mechanism in dynamic VO grids can use platform attestation in addition, or instead of, relying on identity-based credentials. Grid technology will become more feasible for commercial use by incorporating Trusted Computing technology, which allows trust to be established in a manner that matches commercial requirements.

Acknowledgements

This work is being funded by the Engineering and Physical Sciences Research Council (EPSRC) UK e-Science programme of research (EP/D053269). For more details of this project please refer to www.distributedtrust.org.

References

D. Chadwick, A. Otenko, and E. Ball. Role-based access control with x.509 attribute certificates. *IEEE Internet Computing*, 7(2):62–69, March 2003.

A. Cooper and A. Martin. Towards a secure, tamper-proof grid platform. In *Proceedings of the Sixth IEEE International Symposium on Cluster Computing and the Grid, Singapore, May 2006*, pages 373–380. IEEE Press, May 2006.

T. Dierks and C. Allen. The TLS Protocol Version 1.0. RFC 2246, Internet and Engineering Task Force, January 1999.

T. Doyle. Meeting the particle physics computing challenge. *PSCA International Public Service Review: Trade and Industry*, (8):88–89, Autumn 2005.

Enterprise Grid Alliance. Accelerating the adoption of Grid solutions in the enterprise. White paper, Enterprise Grid Alliance, Dec 2004.

I. Foster and C. Kesselman. *The Grid 2: Blueprint for a New Computing Infrastructure*. Morgan Kaufmann Publishers, San Francisco, 2nd edition, 2004.

E. M. Gallery and A. Tomlinson. *Secure delivery of conditional access applications to mobile receivers*, volume 6 of *IEE Professional Applications of Computing*, chapter 7, pages 195–237. IEE Press, London, 1st edition, 2005.

D. Grawrock. *The Intel safer computing initiative: Building blocks for Trusted Computing*. Intel Press, 2006.

W. Mao, A. Martin, H. Jin, and H. Zhang. Innovations for grid security from trusted computing – protocol solutions to sharing of security resource. In *Proceedings of the 14th International Workshop on Security Protocols, Cambridge, UK, March 2006, to appear*. Springer-Verlag LNCS, March 2006.

W. Mao, F. Yan, and C. Chen. Daonity – Grid security with behaviour conformity from trusted computing. In *Proceedings of the first ACM workshop on Scalable Trusted Computing, Alexandria, Virginia, US*, pages 43–46. ACM Press, Nov 2006.

C. J. Mitchell. *Trusted Computing*, volume 6 of *IEE Professional Applications of Computing*. IEE Press, London, 1st edition, 2005. Oasis Standards. *WS-Trust 1.3*, March 2007.

Thompson and Olson et al. *CA-based Trust Model for Grid Authentication and Identity Delegation*. Open Grid Forum, Oct 2002.

Trusted Computing Group. *TCG Trusted Network Connect TNC Architecture for Interoperability Specification Version 1.1 Revision* 2, May 2005.

Trusted Computing Group. *TPM Main Part 1 Design Principles Specification Version 1.2 Revision 94*, March 2006.

S. Tuecke et al. Internet X.509 public key infrastructure (PKI) proxy certificate profile. RFC 3820, Internet and Engineering Task Force, June 2004.

Section 2

Quantitative Approaches for Knowledge Discovery and Decision Support in the Post-Genomic Era

A semantic Grid Infrastructure Enabling Integrated Access and Knowledge Discovery from Multilevel Data in Ppost-Genomic Clinical Trials

Manolis Tsiknakis[1], *Stelios Sfakianakis*[1], *George Potamias*[1], *Giorgos Zacharioudakis*[1] *and Dimitris Kafetzopoulos*[2]

1 Foundation for Research and Technology-Hellas, Institute of Computer Science, Biomedical Informatics Laboratory, (tsiknaki, ssfak, potamia, gzaxar)@ics.forth.gr
2 Foundation for Research and Technology-Hellas, Institute of Molecular Biology and Biotechnology, Laboratory for Post-Genomic TEchnologies, kafetzo@imbb.forth.gr

Abstract. This paper reports on original results of the ACGT integrated project focusing on the design and development of a European Biomedical Grid infrastructure in support of multicentric, post genomic clinical trials on Cancer. The paper provides a presentation of the needs of users involved in post-genomic CTs, and presents such needs in the form of scenarios which drive the requirements engineering phase of the project. Subsequently, the initial architecture specified by the project is presented and its services are classified and discussed. The Master Ontology on Cancer, been developed by the project, is presented and our approach to develop the required metadata registries, which provide semantically rich information about available data and computational services, is provided. Finally, a short discussion of the work lying ahead is included.

1 Introduction

Recent advances in research methods and technologies have resulted in an explosion of information and knowledge about cancers and their treatment. Exciting new research on the molecular mechanisms that control cell growth and differentiation has resulted in a quantum leap in our understanding of the fundamental nature of cancer cells and has suggested valuable new approaches to cancer diagnosis and treatment.

The ability to characterize and understand cancer is growing exponentially based on information from genetic and protein studies, clinical trials, and other research endeavors. The breadth and depth of information already available in the research community at large, present an enormous opportunity for improving our ability to reduce mortality from cancer, improve therapies and meet the demanding individualization of care needs (Yurkewicz 2006).

While these opportunities exist, the lack of a common infrastructure has prevented clinical research institutions from being able to mine and analyze disparate data

sources. This inability to share technologies and data developed by different organisations is therefore severely hampering the research process.
The vision of the Advancing Clinico-Genomic Trials on Cancer (ACGT) project (www.eu-acgt.org) is to contribute to the resolution of these problems by developing a semantically rich grid infrastructure in support of multi-centric, post-genomic clinical trials (CTs), and thus enabling for discoveries in the laboratory to be quickly transferred to the clinical management and treatment of patients (see figure 1).

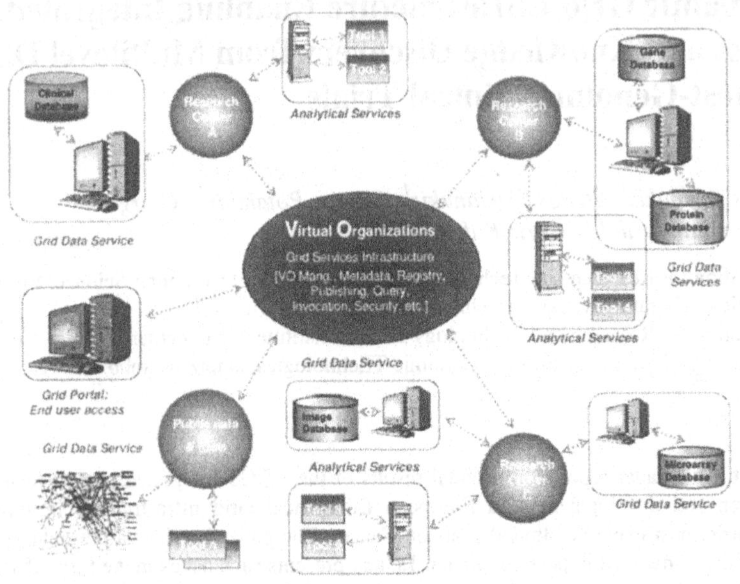

Fig.1: The ACGT semantic grid infrastructure, allowing the creation of dynamic Virtual Organisations (VOs) and the coordinated and secure sharing of data and tools.

The structure of the paper is as follows: Section 2 presents a short background section discussing the urgent needs faced by the biomedical informatics research community; it also presents the clinical trials upon which the ACGT project is based. Section 3 discusses the requirements and presents the architecture defined. Section 4 focuses on two key services of the defined architecture: the Mediator service and the ACGT Master Ontology on Cancer. The issue of Grid intelligence is introduced in section 5 and the need for rich metadata for the description and publishing of Grid resources for enabling their semantic discovery and integration is presented. Finally Section 6 discusses the status and outlook of the work that needs to be completed for the realization of the vision of the project.

2 Background

Biologists and computer scientists are working in designing data structures and in implementing software tools to support biomedicine in decoding the entire human genetic information sequencing. But, many issues are still unsolved. Among the most critical of these are the issues of heterogeneous data integration and metadata definitions (Cannataro 2004).

This need for integration is to some extent clear in the case of complex, multifactorial diseases, such as cancer. Because our knowledge of this domain is still largely rudimentary, investigations are now routinely moving from being "hypothesis-driven" to being "data-driven" with analysis based on a search for biologically relevant patterns. In this context, exploratory analyses is the process of generating hypotheses that are later supported (or not) by the data (e.g. hypothesis: gene x is responsible for a side effect of drug y). The task of validating these hypotheses is done by means of clinical trials.

2.1 The ACGT Post Genomic Clinical Trials

Within such a framework, the ACGT project has been structured. The project has selected two cancer domains and has defined specific trials, which are feeding the requirement analysis and elicitation phase of the project, in the domain of Breast Cancer and Wilm's Tumor (pediatric nephroblastoma). Due to space limitations we will very briefly describe the trial in the domain of Breast Cancer.
Breast cancer is both genetically and histopathologically heterogeneous, and the mechanisms underling breast cancer development remains largely unknown. The ACGT Test of Principle (TOP) study aims to identify biological markers associated with pathological complete response to anthracycline therapy (epirubicin), one of the most active drugs used in breast cancer treatment. To this end, the neoadjuvant approach is very attractive, as it provides an in vivo assessment of treatment sensitivity without affecting adversely survival (Fan 2006).
Supported by "in-vitro" and preliminary "in-vivo" data, this study is designed to test prospectively the value of topo II alpha gene amplification and protein overexpression in predicting the efficacy of anthracyclines.

2.2 Analytical Scenarios to Be Supported

Having defined the clinical studies to be implemented through the use of the ACGT grid infrastructure, we then proceeded to the task of capturing the functional requirements from such an infrastructure. To this extend, a range of scenarios have been developed by the ACGT user community as well as several "technology-driven" scenarios, with the purpose of eliciting requirements and guiding specifications. The most complex of such scenarios, presenting the "analytical requirements" of a researcher testing a hypothesis to explain the characteristics of non-responding patients who were withdrawn from the TOP trial due to adverse reactions to treatment, is presented below.
For the realization of the required analytical tasks, users need to be supported by the platform in executing the following steps, which constitute the "analytical scenario" or the "scientific workflow":

- Query the distributed and heterogeneous clinical trial databases with the purpose of identifying the TOP trial patients' cases with inflammatory breast cancer that show less than 50% tumour regression and chromosomal amplification in region 11q, who received less than 1 Epirubicine cycle due to serious adverse event allergy in the clinical trial databases of all cancer centers participating in the clinical trial.
- Exclude those patient cases that show polymorphisms in the specific

glucuronidating enzyme of epirubicin UGT2B7.
- Query the corresponding genomic databases (microarray data) for the pre-operative and post-operative gene expression data of these patients.
- Normalize the retrieved data, from all genomic databases participating in the trial, using a selected transformation method.
- Compare with the shown differential gene expression between pre-operative and post-operative data and subsequently cluster the identified genes using an appropriate hierarchical clustering method and tool.
- Visualize the 50 most over-expressed and under-expressed genes.
- Obtain functional annotation for each of those genes from the GO and GeneBank public databases, and identify those genes expressed in B-lymphocytes from public biomedical databases.
- Map those genes into regulatory pathways using a selected visualization tool.
- Finally, get the literature related to kinases present in pathway A and Pathway B and identify their regulatory factors.

3 Initial System Architecture

From the preceding description of the type and range of user requirements, it is apparent that a truly complex technical infrastructure needs to be developed, if support for such integrated access, analysis and visualization of multilevel, heterogeneous data is to be provided in the context of discovery-driven exploratory analysis.

A detailed analysis of the scientific and functional requirements of the ACGT infrastructure was performed, together with an analysis of current state-of-the art in terms of technological infrastructure, data resources, data representation and exchange standards, and ontologies.

With respect to the state-of-the-art, on the one hand the myGrid project (www.mygrid.org.uk) focuses in providing support of investigator-driven experiments in silico. On the other hand the cancer Biomedical Informatics Grid (caBIG - https://cabig.nci.nih.gov/) focuses on the creation of a virtual community that shares resources and tackles the key issues of cyber infrastructure.

From the technical point of view, the requirements identified can be met using a federated, multi-layer, service oriented, and ontology driven architecture. The ACGT project decided to build on open software frameworks based on the WS-Resource Framework (WSRF) and Open Grid Service Architecture (OGSA), the de facto standards in Grid computing. These standards are implemented in the middleware selected, namely Globus Toolkit 4 (GT4) (www.globus.org) and Gridge (http://fury.man.poznan.pl/gridge/).

An overview of the ACGT system layered architecture is given in Fig. 2, which is shortly presented in the sequel.

- COMMON GRID INFRASTRUCTURE LAYER: This layer comprises the basic "Grid engine" for accessing remote resources in grid environment. It provides common interface for grid resources used by higher level services.

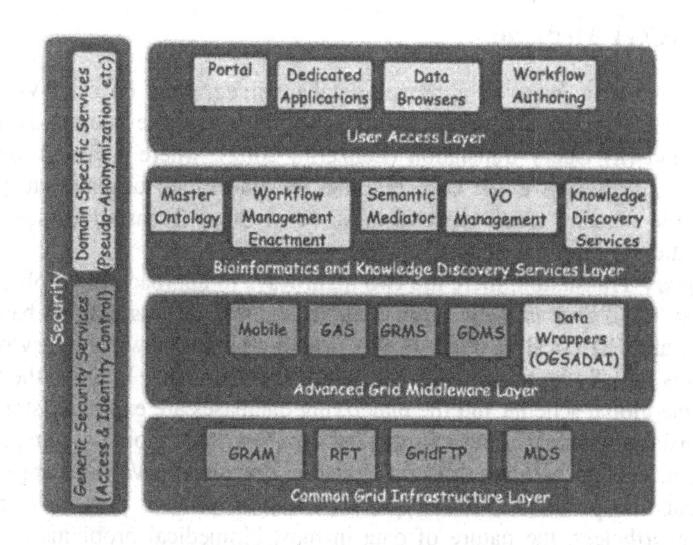

Fig. 2. The ACGT layered functional architecture.

- ADVANCED GRID MIDDLEWARE LAYER: This layer comprises advanced Grid services, which operate on sets of lower level services to provide more advanced functionality.
- BIOINFORMATICS AND KNOWLEDGE DISCOVERY SERVICES LAYER: This layer includes all the ACGT specific services, such as the ACGT Master Ontology, the Clinical Trial on Cancer Metadata Services, the Semantic Mediation services and the Distributed and Privacy preserving Data Mining and Knowledge Discovery services.
- USER ACCESS LAYER: This layer allows users to realize complex biomedical applications as composition of basic services from the underlying layers exploiting the resources and data provided by research centres forming different CT VOs.
- SECURITY LAYER: Access rights, security, and trust buildings are issues addressed and solved on this layer.

4 The Bioinformatics and Knowledge Discovery Services Layer

The Bioinformatics and Knowledge Discovery Services layer includes all ACGT specific services that provide either analytical capabilities or support the integrated knowledge discovery process. Knowledge discovery has been described as "the non-trivial process of identifying valid, novel, potentially useful, and ultimately understandable patterns in data" (Fayyad 1996).

Two core services at this layer are the mediator and the Master Ontology. The former offers integrated query services, while the latter provides the necessary semantic background during the integration process. We will briefly describe these in the following two sub-sections.

4.1 The ACGT Mediator

There are two main ways to approach database integration: (i) Data Warehousing (Kimball, 2002), where a centralized repository represents the integration and stores the data, and (ii) Query translation (Lenzerini 2002), where a virtual view of the integration is presented to the user, but data remain in their original databases. The nature of the domain, make approaches based in Query Translation more adequate for integration of biomedical data sources.
Within Query Translation, there are two main ways to approach the problem of data integration: (i) Global as View (GaV), where a global view is defined based on the databases, and (ii) Local as View (LaV), where a single view for every one of the databases is defined by means a global model of the domain. In GaV, the mappings between the global schema and the underlying databases are explicitly stored, so the query translation process is straightforward. However, this global schema is difficult to maintain. GaV is adequate for non-changing domains. LaV-based approaches do not present this problem, but the query translation process presents performance issues. Nevertheless, the nature of data in most biomedical problems suggests the selection of LaV.
The mediation layer in ACGT is comprised by several components, with the semantic mediator been the main one. It is supported by the ACGT Master Ontology (see next section). The Master Ontology is used by the semantic mediator as a conceptual background to which the schemas of the underlying data sources are mapped, enabling proper semantic integration of data.
The mediator itself acts as a client of other components in the ACGT platform, called "Data Access Services", which offer seamless access to the integrated data sources. For each source type to be included in the platform, such a service is created. This way, the complexities and peculiarities of the sources are hidden under a common interface. The selected query language for the wrappers is SPARQL (http://www.w3.org/TR/rdf-sparql-query/). Since the interface employed in the mediator generates queries in RDQL, a parser for translating RDQL into SPARQL has also been developed.

4.2 The ACGT Master Ontology

As discussed previously, ACGT seeks to provide complex data querying and mediation functionality for the ACGT Grid infrastructure. For this task Master Ontology is required, in order to provide the foundations for semantic data integration, as already discussed in the previous section. The ultimate objective is to provide common domain ontology for the cancer research and management, in order to avoid case-by-case resolutions.

Fig. 3. The BFO system

The ACGT Master Ontology is ontology of cancer research and management with the objective of enabling semantic data integration. The ontology is been build from a realist point of view and will thus, not deal with concepts but rather with universals or classes. The method employed to create the ontology is the application of philosophical and logical principles. Therefore, the ACGT Master Ontology is been developed to include formal is-a, value restrictions, general logical constraints and relations. As a result, the ACGT Master Ontology covers all criteria of ontologies as presented in Grenon, et al (Grenon 2004).
The top level of the ACGT Master Ontology is based on the Basic Formal Ontology (BFO). A central feature of BFO is the basic dichotomy between continuants and occurents which emphasizes two distinct modes of existence in time which is sometimes referred to as the SNAP-SPAN ontology (Zhong 2004). Continuants are entities that exist over time, they are objects of change (e.g. human being, tumor, molecule), whereas occurents are entities existing in time, they are change itself (e.g. breathing, growth, cell division). Furthermore BFO is currently available in an

OWL-DL implementation which increases the possibilities of syntactical integration and reasoning.
Choosing a coherent and logically consistent top level for the ontology is a highly important step. This "Top Down" part of ontology development is vitally important in order to come to common terms and principles.
The first step of adding entities from clinical practice in the ontology was the integration of clinical report forms (CRF) into the system. The CRFs represent data from the different data types in the ACGT domain, except, in the most cases, molecular data. In an approach that could be called a "Bottom Up", universals were edited to which patient data refers. The development of the ACGT Master Ontology started with the trials selected, but it is easy to extend its scope, once the first steps in categorization are taken. The existence of a coherent top level ensures the reusability of the ontology, since it prevents the development of ontologies based on a top level which is restricted to one specific domain.

5 Biomedical Grid Intelligence

The way how data at different levels of the Grid can be effectively acquired, represented, exchanged, integrated and converted into useful knowledge is an emerging research field known as "Grid Intelligence". The term indicates the convergence of web service, grid and semantic web technologies and in particular the use of ontologies and metadata as basic elements through which intelligent Grid services can be developed. An example of this convergence is the Semantic Grid (d. Roure 2003) which focuses on the systematic adoption of metadata and ontologies to describe Grid resources, to enhance and automate service discovery and negotiation, application composition, information extraction, and knowledge discovery, whereas Knowledge Grids (Cannataro 2003) offer high-level tools and techniques for the distributed mining and extraction of knowledge from data repositories available on the Grid, leveraging semantic descriptions of components and data, as provided by Semantic Grid, and offering knowledge discovery services.
In a "grid enabled" data sharing VO, datasets may not be well known amongst all participants of the VO. To integrate the highly fragmented and isolated data sources, we need semantics to answer higher-level questions. Therefore, it becomes critically important to describe the context that the data was captured. We describe this contextualization of the data as "metadata" (data about data). Semantic integration, therefore, in ACGT relies on metadata publishing and ontologies.
A similar approach is reported by other initiatives. The myGrid project relies heavily on the use of semantic descriptions to allow more precise searching by both people and machines and workflows (Wroe 2003).
We see our main future research challenges in ACGT the requirement to develop an infrastructure able to produce, use and deploy knowledge as a basic element of advanced applications, which will mainly constitute a Biomedical Knowledge Grid. In achieving such an objective metadata is critical.
We use OWL-S for developing metadata and service ontologies for describing Grid Services so that they might be discovered, explained, composed and executed automatically. Our initial investigations have also revealed the need for a sophisticated model of provenance, since the use of elementary workflows as well as advanced workflows, i.e. workflows containing other workflows, is becoming a very

important goal in our R&D work. This requirement is also pushing for complex metadata about workflows to be maintained in the ACGT Grid middleware.

6 Conclusions

In this paper we consider a world where biomedical software modules and data can be detected and composed to define problem-dependent applications. We wish to provide an environment allowing biomedical researchers to search and compose bioinformatics and other analytical software tools for solving biomedical problems. We focus on semantic modelling of the goals and requirements of such applications using ontologies.
The infrastructure been developed uses a common set of services and service registrations for the entire clinical trial on cancer community. The shared ACGT semantic services provide biomedical ontologies in common use across clinical trials and cancer research.
The project is in its initial phases of implementation with more than three years of implementation remaining. We are currently focusing in the development of the core set of components up to a stage where they can effectively support in silico investigation and initial prototypes have been useful in crystallizing requirements for semantics within e-Science. The selected demonstrators, stemming from the user defined scenarios, together with these core set of components will enable us to both begin evaluation and gather additional and more concrete requirements from our users. These will allow us to improve and refine the initial architecture and its services.

Acknowledgments

The authors would like to thank all members of the ACGT consortium. The ACGT is partly funded by the EC and the authors are grateful for this support.

References

Cannataro, M. and Talia, D., (2003), *KNOWLEDGE Grid - An Architecture for Distributed Knowledge Discovery*, CACM, vol. 46, no 1, pp. 89-93.

Cannataro, M., et al. (2004), Proteus, a Grid based Problem Solving Environment for Bioinformatics: Architecture and Experiments, IEEE Computational Intelligence Bulletin, vol.3, no.1, 7-18.

d. Roure, D., et al., (2003), *The Semantic Grid: A future e-Science infrastructure*, In: *Grid Computing: Making The Global Infrastructure a Reality*, (Eds.) Berman, F., Hey, A. J. G. e Fox, G., John Wiley & Sons, pp. 437-470.

Fan, C. et al., (2006), Concordance among Gene-expression-based predictors for breast cancer, NEJM, vol. 355, 560-569.

Fayyad, U., Piatetsky-Shapiro, G., Smyth, P., (1996), *From Data Mining to Knowledge Discovery: An Overview.* In Advances in Knowledge Discovery and Data Mining, eds. U. Fayyad, G. Piatetsky-Shapiro, P. Smyth, and R. Uthurusamy, Menlo Park, Calif.: AAAI Press, pp. 1-30.

Grenon, P., Smith, B. and Goldberg L., (2004), *Biodynamic Ontology: Applying BFO in the Biomedical Domain*, in: Ontologies in Medicine, D. M. Pisanelli, Ed., Amsterdam: IOS Press, pp. 20-38.

Kimball, R., (2002), *The Data Warehouse Toolkit: The Complete Guide to Dimensional Modeling* (Second Edition), Wiley, 2002. ISBN 0-471-20024-7

Lenzerini. M., (2002), *Data Integration: A Theoretical Perspective.* In PODS, pp. 233–246.

Wroe, C., et al, (2003), *A suite of DAML+OIL ontologies to describe bioinformatics web services and data*, International Journal of Cooperative Information Systems, vol. 12, no. 2, pp. 197–224, 2003.

Yurkewicz, K., (2006), *Accelerating Cancer Research*, Science Grid, June 21, 2006. Available at http://www.interactions.org/sgtw/2006/0621/cabig_more.html.

Zhong, N. and Liu, J. (Eds.), (2004) *Intelligent Technologies for Information Analysis*, Springer Verlag, ISBN 3540406778.

Enhancement of Multispectral Chromosome Image Classification Using Vector Median Filtering

Petros S. Karvelis[1], Dimitrios I. Fotiadis[1]

1 Unit of Medical Technology and Intelligent Information Systems, Dept. of Computer Science, University of Ioannina, Ioannina, Greece, GR 45110
{pkarvel, fotiadis}@cs.uoi.gr

Abstract. Multiplex in-situ hybridization (M-FISH) is a combinatorial labeling technique in which each chromosome is labeled with 5 fluors and a DNA stain and is used for chromosome analysis. Although M-FISH facilitates the visual detection of gross anomalies, misclassified pixels and cross-hybridization often makes manual examination difficult and introduces operator bias. The success of the technique largely depends on the accuracy of pixel classification. In this work we study the use of nonlinear Vector Median Filtering (VMF) methods to induce the accuracy of pixel classification. We have evaluated our methodology using a subset of images publicly available and the classifier was trained and tested on non-overlapping chromosome images. An overall accuracy of 74.13% is reported when introducing VMF.
Keywords: M-FISH, Watershed segmentation, Vector Median Filtering.

1 Introduction

Chromosomes are the condensed form of the genetic material and their images taken during cell division are useful for diagnosing genetic disorders and for studying cancer [1]. Cytogenetics, study the hereditary material at the cellular level and since the development of M-FISH technique [2] the resolution of detection of genetic abnormalities in cancer has been enhanced significantly.

M-FISH paints all 24 types of chromosomes in a human cell using different colors. This is accomplished using 5 fluors and a fluorescent DNA stain called DAPI (4',6-Diamidino-2-phenylindole). The other five fluors attach to specific sequences of DNA in a way that each class of chromosome absorbs a unique combination of fluors. At least 5 fluors are needed for combinatorial labeling to uniquely identify all 24 chromosomes. In comparison with conventional grayscale chromosome analysis [3], superior detection of subtle and complex chromosomal rearrangements can be obtained by M-FISH analysis [4].

A typical M-FISH set consists of 5 channel images labeled with the dyes Aqua, Far Red, Green, Red, and Gold. From these 5 channel images, the human chromosomes are classified into 24 classes, displayed in different colors. Fig.1 shows such an example of an M-FISH image.

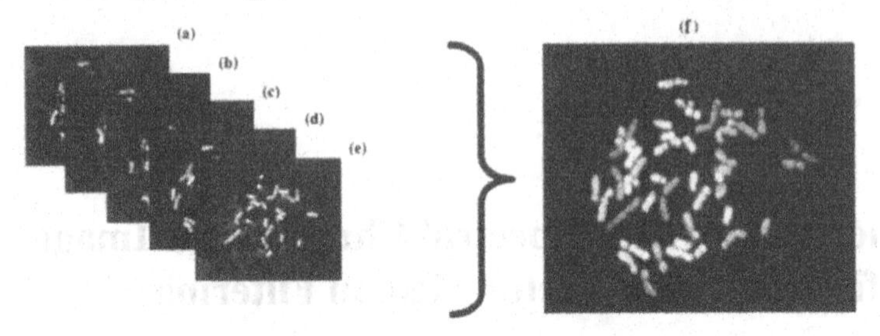

Fig. 1. Five channel M-FISH image data: (a) Aqua fluor, (b) Far red fluor, (c) Green fluor, (d) Red fluor, (e) Gold fluor and (f) Color M-FISH image.

Many attempts have been made to automate parts of the M-FISH chromosome image analysis. Despite their success, they report less than perfect classification accuracy [5]. Indeed, the size of misclassified regions is significant compared to those of the small regions involved in the complex chromosome abnormalities [6]. To make this technique practical for identifying chromosome abnormalities in cancer and genetic disease diagnosis, the key step is to increase the classification accuracy.
In this work we have studied several multichannel image filtering approaches. Especially the vector median filtering and its variation are studied and validated thoroughly using M-FISH images. Vector filters represent a natural approach to the noise removal in multichannel images, since these filters utilize the correlation between color channels. Hence, the vector methods represent the optimal and attractive approach to denoise multichannel images. By performing multichannel filtering, accuracy of the pixel-by-pixel technique is improved. It is reasonable to believe that the improved accuracy of the classifier brought about by these filtering approaches would result in more accurate detection of subtle genetic rearrangements for cancer diagnosis and research.

1.1 Related Work

Several methods have been proposed for M-FISH image classification [6-10]. Most of these methods are based on pixel-by-pixel classification and tackle classification as a problem with 6-features (six channels) and 24-categories. Although the performance of these methods is very promising (accuracy ~90%) for some set of images [6-10], the average pixel classification accuracy for the whole set (~200 images) was only 68% ± 17.5 [8].
Over the last decade, a variety of nonlinear vector filtering approaches for color image filtering has been proposed [11-17]. It has been widely recognized that the processing of color image data as vector fields is desirable due to the correlation that exists among the image channels and to the fact that the nonlinear vector processing of color images is the most effective way to filter out noise.
The most popular nonlinear, multichannel filters are based on the ordering of vectors in a predefined sliding window. The output of these filters is defined as the lowest ranked vector according to a specific ordering technique. The most well-known vector filters include the Vector Median Filter (VMF) [11], the Basic Vector

Directional-Distance Filter (BVDF) [12] and the Directional Distance Filter (DDF) [13]. These nonlinear filters, based on the ordering operation, provide robust estimation in environments corrupted by impulsive noise and outliers. On the other hand, weighted nonlinear vector filtering techniques have been proposed to achieve better performance in noise suppression and detail preservation, which include Weighted Vector Median Filters (WVMF) [14,15]. Finally, Center-Weighted Vector Median (CWVM) filter [16,17] is a special type of the WVM filter where only the weight of the central pixel is adjustable.

2 Multichannel Filtering Schemes

In multichannel image filtering, each image pixel can be considered as a vector of components associated with the intensities of the channels. Thus, it is necessary to consider the correlation that exists among the channels and to apply the vector processing. If the existing correlation is not taken into account and channels are processed independently, then the filtering operation is applied component wise.
In general, component wise (marginal) approaches produce new vector samples, i.e., color artefacts, caused by the composition of reordered channel samples. Vector filters represent a natural approach to the noise removal in multichannel images, since these filters utilize the correlation between color channels. The output of these filters is defined as the lowest ranked vector according to a specific ordering technique.
Let $I(x,y):\mathbb{R}^2 \rightarrow \mathbb{R}^m$ represent a multichannel image, where m characterizes the number of channels. Suppose a square filter window with a set of input multichannel samples such that $X=\{x_i : i=1,2,\ldots,N\}$, where $x_i \in \mathbb{R}^m$ and N is an odd integer, which represents the size of the window.
Let us consider an input sample $x_i : 1 \le i \le N$, associated with the distance measure L_i and A_i defined as:

$$L_i = \sum_{j=1}^{N} \left\| x_i - x_j \right\|_\gamma , \tag{1}$$

$$A_i = \sum_{j=1}^{N} \cos^{-1}\left(\frac{x_i x_j^T}{|x_i||x_j|} \right), \tag{2}$$

where γ characterizes the employed norm, and $|\cdot|$ the magnitude of the vector. Note that the well-known Euclidean distance corresponds to $\gamma = 2$.
The ordering criterion could be expressed through products:

$$\Omega_i = L_i A_i = \sum_{j=1}^{N} \left\| x_i - x_j \right\|_\gamma \sum_{j=1}^{N} \cos^{-1}\left(\frac{x_i x_j^T}{|x_i||x_j|} \right), \; 1 \le i \le N \; . \tag{3}$$

Then, the ordered set is given by, $\Omega_1 \le \Omega_2 \le \ldots \le \Omega_N$. The same ordering scheme applied to the input set results in the ordered sequence, $x^{(\Omega_1)} \le x^{(\Omega_2)} \le \ldots \le x^{(\Omega_N)}$. The sample $x^{(\Omega_1)}$ associated with Ω_1 represents the output of the directional distance filter (DDF). Let us assume the DDF with the power parameter p so that the power

$1-p$ is associated with the sum of vector distances and the power $p \in [0,1]$ is associated with the sum of vector angles. Thus, Eq. (3) can be simply rewritten as:

$$\Omega_i = L_i^{1-p} A_i^p = \left(\sum_{j=1}^{N} \left\| x_i - x_j \right\|_\gamma \right)^{1-p} \left(\sum_{j=1}^{N} \cos^{-1} \left(\frac{x_i x_j^T}{|x_i||x_j|} \right) \right)^p, 1 \le i \le N. \quad (4)$$

If $p=0$, the DDF operates as the vector median filter (VMF), whereas for $p=1$, the DDF is equivalent to the basic vector directional filter (BVDF). The weighted vector median filter is defined through a set of weights. Assume a set of nonnegative integer weights $w_1, w_2, \ldots, w_N$ so that each weight $w_j, 1 \le j \le N$ is associated to each input sample x_j. Then, it is possible to express the weighted vector distance J_i as:

$$J_i = \sum_{j=1}^{N} w_j \left\| x_i - x_j \right\|_\gamma, 1 \le i \le N. \quad (5)$$

The sample $x^{(J_1)} \in \{x_1, x_2, \ldots, x_N\}$ associated with the minimal combined weighted distance J_1 is the sample which minimizes the sum of weighted vector distances and the output of the WVFM filter.

The CWVMF [20, 21] framework is more adequate for an adaptive filter design that will vary the smoothing levels in the filtering process, than the WVMF with a full set of weight coefficients. Consider the weight vector given by:

$$w_j = \begin{Bmatrix} N-2k+2, j=(N+1)/2 \\ 1, otherrwise \end{Bmatrix}. \quad (6)$$

where $k = 1, \ldots, (N+1)/2$. Only the central weight $w_{(N+1)/2}$ associated with the central sample $x_{(N+1)/2}$ can be changed, whereas other weights associated with the neighboring samples remain equal to one. If the smoothing parameter k is equal to one, then the CWVMF is equivalent to the identity operation and no smoothing will be performed. In the case $k=(N+1)/2$, the maximum amount of smoothing will be performed and the CWVMF filter is equivalent to the VMF.

3 Pixel-by-Pixel classification

We used the classification of each pixel, to measure and validate the performance of the filtering schemes because the ultimate goal of these approaches is to improve the pixel classification accuracy. For this purpose, we employed the widely used Bayesian classifier [7]. Each pixel in the metaphase chromosome image corresponds to a vector Y with five elements:

$$Y = (y_1, y_2, y_3, y_4, y_5). \quad (7)$$

where each element y_i of the vector represents the gray level value in one of the five image channels. To classify a pixel described by the feature vector Y, we calculate the a posteriori probability $P(c_i \mid Y)$, that the pixel belongs to class c_i:

$$P(c_i \mid Y) = \frac{p(Y \mid c_i)p(c_i)}{P(Y)}, \; P(Y) = \sum_{i=1}^{24} p(Y \mid c_i)p(c_i). \tag{8}$$

where $P(c_i),\ i = 1,2,\ldots,24$ is the a priori probability that a feature belongs to class c_i and $p(Y \mid c_i)$ denotes the class conditional probability distribution function. This represents the probability distribution function, for a feature vector Y given that Y belongs to class c_i.

Given the nature of data the probability distribution of the feature values for each class c_i is assumed to be multivariate normal density function, with a probability density function:

$$p(Y \mid c_i) = \frac{1}{(2\pi)^{5/2}\left|\Sigma_i\right|^{1/2}} \exp\left[-\frac{1}{2}(Y-\mu_i)^T \Sigma_i^{-1} (Y-\mu_i)\right]. \tag{9}$$

where μ_i is the mean vector and Σ_i is the 5×5 covariance matrix of the class c_i, $\left|\Sigma_i\right|$ and Σ_i^{-1} are the determinant and the inverse, respectively.

4 Results

In order to compare the performances of the applied filtering schemes, we used a set of M-FISH images from the ADIR database [18]. The database contains six-channel image sets recorded at different wavelengths. Each dataset includes a "ground truth" image for each M-FISH image in which each pixel is labeled according to the class to which it belongs. This image is labeled such that the gray level of each pixel is its class (chromosome) number. Background pixels are 0 and pixels in a region of overlap are -1. We used the ground truth segmentation in order to assess the performance of the different multichannel filtering schemes. 17 were randomly chosen, from which two were used for training and the remaining for testing. The performance was measured in terms of accuracy. The classification results for the various multichannel filtering schemes are shown in Table I.

Table1. Comparison of different filtering schemes

	Without Filter	VMF	BVDF	CWVMF
Overall (%)	69.98	73.51	74.13	74.03

5 Conclusions

The pixel misclassification errors in M-FISH images result from a number of factors, including biochemical noise, electronic noise, and spectral overlap. An image filtering step is required for improving accuracy in pixel classification. Improved classification accuracy will greatly decrease the size and number of misclassified pixel regions. This would allow smaller rearrangements to be identified and better enable the technique to resolve complex rearrangements when applied to patient diagnosis. In this work the effect of multichannel vector filtering on M-FISH images was studied. Vector median filtering and its variations were proposed in detail and proposed for the enhancement of multichannel chromosome image classification. As it can be seen from Table I, BVDF achieves the best results in terms of accuracy improving the classification accuracy by 4.15% while CWVMF attains the worst improvement by 1.54%. The enhancement algorithms have achieved objective improvement with medical significance in terms of the improvement of classification of chromosomes.

References

Thompson, M., McInnes, R., Willard, H.: Genetics in Medicine. 5th Edition, WB Saunders Company, (1991) Philadelphia.

Speicher, M.R., Ballard, S.G., Ward, D.C.: Karyotyping human chromosomes by combinatorial Multi-Fluor FISH. Nat. Genetics, 12. (1996) 341-344.

Wang, Y.P., Wu, Q., Castleman, K.R., Xiong, Z.: Chromosome image enhancement using multiscale differential operators. IEEE Trans. Med. Imag. 22 (2003) 685-693.

Veldman, T., Vignon, C., Schröck, E., Rowley, J.D., Ried, T.: Hidden chromosome abnormalities in hematological malignancies detected by multicolor spectral karyotyping. Nat. Genetics. 15 (1997) 406–410.

Castleman, K.R., Morrison, E., Piper, J., Saracoglu, K., Schultz, M., Speicher, M.R.: Classification accuracy in multiple color fluorescence imaging microscopy. Cytometry. 41 (2000) 139–147.

Wang, Y., Castleman, K.R.: Normalization of multicolor fluorescence in situ hybridization (M-FISH) images for improving color karyotyping. Cytometry. 64 (2005) 101-109.

Sampat, M.P., Castleman, K.R., Bovik, A.C.: Pixel-by-Pixel classification of MFISH images. In Proc. 24th Annual International Conference of the IEEE EMBS. (2002) 999-1000.

Schwartzkopf, W.C., Bovik, A.C., Evans, B.L.: Maximum-likelihood techniques for joint segmentation-classification of multispectral chromosome images. IEEE Trans. Med. Imag. 24 (2005) 1593-1610.

Sampat, M.P., Bovik, A.C., Aggarwal, J.K., Castleman, K.R.: Supervised parametric and non-parametric classification of chromosome images. Pattern Recognit. 38 (2005) 1209-1223.

Karvelis, P.S., Fotiadis, D.I., Georgiou, I., Syrrou, M.: A watershed based segmentation method for multispectral chromosome image classification. In Proc. 28th Annual International Conference of the IEEE EMBS. (2006) 3009-3012.

Astola, J., Haavisto, P., Neuov, Y.: Vector median filter. Proc. IEEE. 78 (1990) 678-689.

Trahanias, P.E., Venetsanopoulos, A.N.: Vector directional filters—a new class of multichannel image processing filters. IEEE Trans. Image Process. 2 (1993) 528–534.

Trahanias, P.E., Karakos, D. Venetsanopoulos A.N.: Directional processing of color images: Theory and experimental results. IEEE Trans. Image Process. 5 (1996) 868–881.

Viero, T., Oistamo, K., Neuvo, Y.: Three-dimensional median related filters for color image sequence filtering. IEEE Trans. Circ. Syst. Video Technol. 4 (1994) 129–142.

Alparone, L., Barni, M., Bartolini, F., Caldelli, R.: Regularization of optic flow estimates by means of weighted vector median filtering. IEEE Trans. Image Process. 8 (1999) 1462–1467.

Lukac, R.: Adaptive color image filtering based on center-weighted vector directional filters. Multidimensional Syst. Signal Process. 15 (2004) 169–196.

Lukac, R., Smolka, B., Plataniotis, K.N., Venetsanopoulos, A.N.: Selection weighted vector directional filter. Comput. Vis. Image Understand. 94 (2004) 140–167.

The ADIR M_FISH Image Database. Available at: http://www.adires.com/05/Project/MFISH_DB/MFISH_DB.shtml.

Extending the Interpretation of Gene Profiling Microarray Experiments to Pathway Analysis Through the Use of Gene Ontology Terms

Aristotelis Chatziioannou[1], *Panagiotis Moulos*[1]

1 Insitute of Biological Research and Biotechnology, National Hellenic Research Foundation, 48 Vassileos Constantinou ave., 11635 Athens, Greece
{achatzi, pmoulos, kolisis}@eie.gr

Abstract. Microarray technology allows the survey of gene expression at a global level by measuring mRNA abundance. However, the grand complexity characterizing a microarray experiment entails the development of computationally powerful tools apt for probing the biological problem studied. Here we propose a suite for flexible, adaptable to a wide range of possible needs of the biological end-user, data-driven interpretation of microarray experiments. The suite is implemented in MATLAB and is making use of two modules, able to perform all steps of typical microarray data analysis starting from data standardization and normalization up to statistical selection and pathway analysis utilizing Gene Ontology Term annotations for the species genomes interrogated, whereas due to its modular structure it is scalable thus enabling the incorporation or its seamless assembly with other existing tools.

Keywords: microarray data analysis, noise filtering, normalization, statistical selection, clustering, pathway, gene ontology analysis.

1 Introduction

Microarray experiments are high throughput measuring techniques monitoring simultaneously the gene expression of the whole genome of the species studied, across a distinct set of conditions. The advent of microarrays has been proven a major discovery breakthrough in biological research with a deluge of studies coming out lately. Using microarrays for diagnostic purposes, entails the discovery of gene-sets, whose expression pattern consistently distinguishes among disease and physiological states, functioning as genetic signatures for the disease probed. In this sense, gene expression profiling can be used for the scope of: i) identifying and categorizing diagnostic or prognostic biomarkers ii) classifying diseases, e.g. tumours with different prognosis not distinguishable by classical histological assays iii) monitoring of therapeutic efficacy and iv) understanding the mechanisms involved in the genesis and progression of disease [1].

In a typical gene profiling experiment, mRNA is extracted from tissues or cells, reversed-transcribed, labelled and hybridized onto the array. Washing is following under a stringent protocol to diminish the possibility of non-specific hybridization. The next step is image acquisition and segmentation where quantitative estimates of the relative fluorescence intensities of each spot are obtained [1], [2]. The derived images are used to create a dataset which needs pre-processing in order to correct for several systematic measurement errors. Typical pre-processing steps are background noise correction to adjust for non-specific hybridization, presence of array artefacts, washing issues or quantum fluctuations, filtering procedures to eliminate non-informative genes, calculation of the logarithmic transformed ratio between Cy5 and Cy3 channels and data normalization. There exist several methods for normalizing cDNA microarray data [3], [4] and abundant literature is available on the subjects of pre-processing and normalization [2], [5], [6]. A major drawback, concerning the several commercial or open source microarray data analysis software implementations, is the lack of a standardized, consistent in the sense of default pre-selected methods, analysis pipeline yet flexible, if other selections are to be made for the various stages of data processing. A practical result of this would be, effective batch processing that starts from raw image analysis data and results in sound lists of differentially expressed (DE) genes.

No matter how important the role of statistical gene selection might be, the extraction of DE gene lists constitutes still a primary step often offering more questions than answers, regarding the biological interpretation of the problem probed and the mechanisms underlying it. On the other hand, these lists constitute a convenient entry point for tracing the biological pathways involved in the mechanism investigated. This process is named 'meta-analysis'. Among others, meta-analysis includes pathway analysis in the sense of uncovering genes with a certain expression profile mapped to the same pathway, exploration for common regulatory elements among groups of genes, and gene functional analysis based on biological databases or ontologies. The Gene Ontology (GO) database [7] provides such functional annotation in a hierarchical way constituting a valuable resource for meta-analysis.

Many meta-analysis software tools are currently available, commercial or freely accessible (e.g. GenMAPP, [8], GOALIE, [9] among others). Some simply offer aid by visualizing pathways and mapping the respective genes and annotation while others use sophisticated statistical or graph-theoretical techniques to mine important biological functions in a hierarchical fashion. The suite presented in this paper, constitutes a free and open source stable microarray data analysis platform implemented in MATLAB, expanding the capabilities of ANDROMEDA (Automated aND RObust Microarray Experiment Data Analysis, [10]) so as to support the crucial task of ranked statistical pathway extraction by exploiting gene annotations according to the GO repository. Its power relies in user-friendliness, flexibility, modularity and automation which constitute helpful features for the common experimental biologist.

2 The Software

ANDROMEDA is a pipeline programmed in MATLAB for interpretation of microarray experiments, which provides a unified environment for batch processing, starting from reading raw image analysis software output data and resulting in lists of DE genes and hierarchical gene clusters. The analysis can be fully automated or adjusted according to user's needs by supporting a wide repertoire of functions at different stages. The program supports numerous filtering, normalization and statistical methods to scrutinize the data for any number of experimental conditions. An advantage of ANDROMEDA, is that while most tools specialize either in visualization and normalization, or statistical testing and clustering, all these steps are implemented in a pre-defined workflow [10], thus uniting the whole analysis process, and bearing a great benefit in terms of time efficiency just as the avoidance for the support of numerous data-conversion formats.

Following the import of image signal intensity files (currently 4 formats supported namely Genepix, QuantArray, ImaGene, plus a custom generic tab-delimited format for use with other signal intensity formats), spot signal is modified by applying either background subtraction, or calculation of the signal-to-noise ratio or no correction. Background subtraction yields a net signal spot value by correcting for various background effects. However, due to various sources of measurement error [10], [11], [12], different ways of background correction might be applied. This might prove critical, especially when dealing with weak signal datasets, whereas a majority of spot signals is close to background and the possible role of low abundance genes should not be overlooked. Nevertheless, as the background correction issue [12] raises debates, there are analysis pipelines, which use the initial uncorrected spot values.

Concerning poor quality spot filtering, apart from spots flagged as poor automatically or manually, a series of filters are provided for further assessment. One simple filter is a signal-to-noise ratio threshold: spots below that threshold are viewed as noise contaminated and therefore unreliable. Yet, such a filter might prove extremely strict, especially for low intensity genes. Another option is to check the extent of overlap between the signal and background distributions for each spot. The rational behind this filter is that the signal of robust spots should lie pretty far from the background, thus the distance between the medians of the two distributions should be adequate in terms of their respective standard deviations, so as to avoid overlap. Alternatively, possible filter customization is feasible, considering particularities of biological subjects under investigation. Outlier detection is conducted through a t-test (parametric) or Wilcoxon (non-parametric) test which examines each spot, so that its replicate ratio measurements for every condition follow a normal (or a continuous symmetrical) distribution. Array spots sensitive to any of the filters are excluded from the estimation of the normalization curve, to alleviate the normalization procedure from the impact of systematic measurement errors [10].

Data normalization is performed on each slide separately (Figure 1a-c). Subgrid normalization is possible, to allow considering spatial dependent particularities. Given the impact of normalization methods on subsequent analysis steps [13], the decision on which normalization method is proper may depend on the biological nature of the dataset interrogated. Lowess/Loess methods are usually preferred because of intensity dependent data smoothing which can remove systematic biases and intensity dependent effects in the log space and elevate or uncover gene specific biological outcomes. Moreover, the robust versions of Lowess/Loess [4] perform additional fitting iterations over the dataset while readjusting each point's weight on each pass, so as to mitigate the impact of possible outliers.

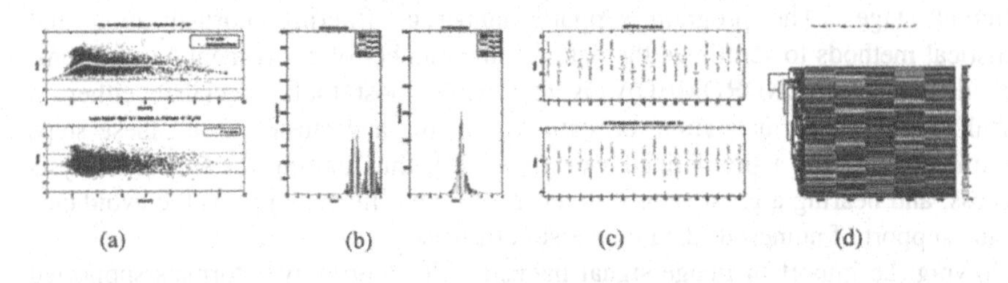

Fig. 1. (a) MA plot before and after the application of robust loess normalization with neighbouring span 0.1 (b) Intensity ratio distributions for an experimental condition with 4 replicates. The normalization impact is profound (c) Boxplots before and after the application or robust loess normalization for a set of 19 slides (d) Hierarchical clustering with average linkage and Euclidean distance for a set of 219 DE genes among 5 experimental conditions.

The statistical selection of DE genes is based on well established and widely used statistical methods [14], [15]. The workflow of statistical selection is analytically described and justified in [10] and seeks to maximize information concerning expression fold-change while attempting to standardize signal variation among microarray slides. Multiple testing correction [16], [17], [18] is performed by controlling the False Discovery Rate [17]. Finally, the obtained DE gene lists are subjective to hierarchical clustering so as to reveal groups of co-regulated genes and assemble gene clusters possibly sharing similar regulatory elements or the same pathway. The main analysis workflow of ANDROMEDA is presented in Figure 2.

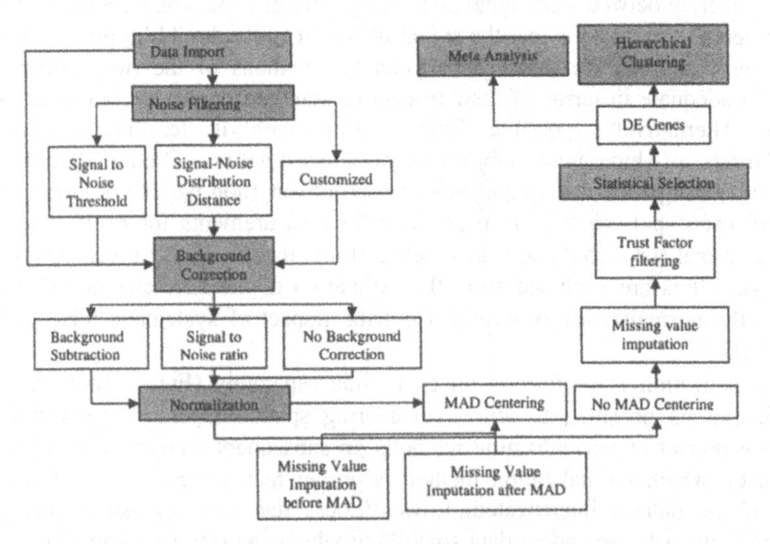

Fig. 2. The ANDROMEDA protocol workflow (see also [10])

The lists of DE genes represent only a first step towards a meaningful functional analysis of a biological problem investigated. To gain further insights over specific biological questions interrogated through a microarray experiment, additional analysis is required exploiting available knowledge databases. GO database is a valuable repository, offering a hierarchical structure of functional terms associated to several groups of genes, reflecting partial or whole biochemical pathways. Using GO together with established statistical methods, the identified significant genes can be related with important nodes in the GO tree structure and cellular actions can be seen as conceptual entities mapped as nodes to an organizational schema.
One way to incorporate results of a microarray experiment to statistically significant molecular functions and pathways is the exploitation of a crucial attribute of the GO Terms (GOTs) tree structure and population derived by each DE gene list; many genes which are hierarchically lower in the context of several biological functions are represented as 'leaves' in the GO tree but are connected to hierarchically higher biological entities through the same structure, inheriting these GOTs too. The main goal of a meta-analysis algorithm would be to determine among all GOTs associated with the significant gene list, those associated with nodes higher in the GO hierarchy, and rank them by a statistical score. In this sense, possible errors in the statistical selection stage of a microarray analysis are significantly mitigated, facilitating the targeting of particular biological objects for further investigation.

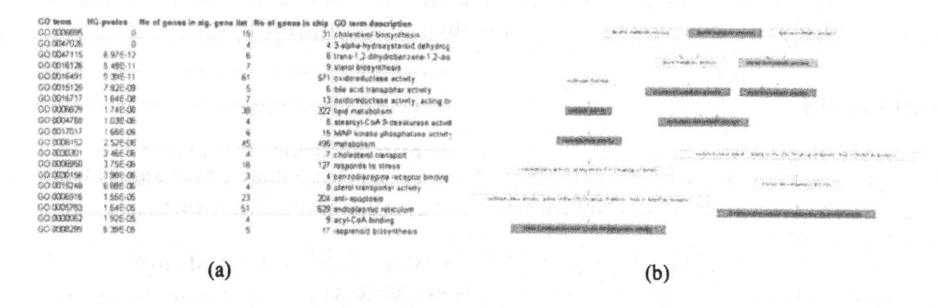

Fig. 3. (a) Short output of pathway analysis module (b) Tree diagram depicting the top 5 most significant GO terms and their ancestors at depth 2. Nodes appearing in the final output list are coloured with different tones of red according to significance.

The significance of GOTs can be assessed with proper statistical tests such as the hypergeometric test, the Chi test or the Fisher exact test [18], using the GO annotation of the significant gene lists and the list of the entire array. The pipeline described extends its capabilities by adopting a set of proprietary functions programmed in MATLAB which implement an algorithm which based on the aforementioned essential properties of GOTs, performs meaningful meta-analysis, yielding a list of significantly deregulated GO functions, by assigning proper statistical scoring through the use of resampling techniques. Data resampling is widely utilized [19], [20] for non-parametric estimation of null distributions used for p-value scoring of proper statistics reflecting properties of the problem analyzed. Such statistics can be empirical, distribution comparison statistics (e.g. Kolmogorov-Smirnov or Anderson-Darling statistics [21]) or modifications of the latter [19]. The results of this algorithm propose complete or parts of molecular pathways, integrating groups of significant genes, rather than isolated ones and statistically sound and biologically relevant functional categories as specific targets for additional examination (Figure 3).

3 Conclusions

The microarray analysis pipeline presented here constitutes a completely open source and flexible microarray data analysis suite, starting from raw image analysis files up to specific biological entities comprising targets for further research on the initial biological issue that triggered the experiment. To our knowledge there are no tools up to now, performing the analysis so extensively and in such a unified way, enabling time saving, user-friendly, effective, versatile in terms of methods, batch processing of microarray experiments. The derived data lists provide concise, yet in depth extensive information concerning either individual genes or gene sets mapped to specific pathways, constituting in this way an ideal starting point for further analysis.
The scalability of the pipeline presented in this work, permits the support of a rich set of further enhancements as future innovations. Indicatively, the parallelization of the platform so as to run in a Grid environment, exploding thus its computational capabilities is already under implementation, whereas another improvement is the development of tools permitting the integration of an as detailed as possible annotated information, in the form of accession numbers referring to numerous biological databases, concerning either individual genes or pathways.

References

Tarca, A.L., Romero, R., Draghici, S.: Analysis of Microarray Experiments of Gene Expression Profiling. American Journal of Obstetrics and Gynecology 195 (2006) 373-388

Quackenbush, J.: Microarray Data Normalization and Transformation. Nat. Genetics 32 (2002) 496-501

Tseng, G.C., Oh, M.K., Rohlin, L., Liao, J.C., Wong, W.H.: Issues in cDNA Microarray Analysis: Quality Filtering, Channel Normalization, Models of Variations and Assessment of Gene Effects. Nucleic Acids Research 29 (2001) 2549-2557

Cleveland, W.S., Grosse, E., Shyu, W.M.: Local Regression Models. In: Chambers, J.M., Hastie, T.J. (eds): Statistical Models in S. Wadsworth & Brooks/Cole Dormand, J.R. (1992)

Yang, Y.H., Dudoit, S., Luu, P., Lin, D.M., Peng, V., Ngai, J., and Speed, T.P.: Normalization for cDNA Microarray Data: a Robust Composite Method Addressing Single and Multiple Slide Systematic Variation. Nucleic Acids Res 30: e15, 2002.

Cui, X., Kerr, M.K., Churchill, G.A.: Transformations for cDNA Microarray Data. Stat. Appl. Genet. Mol. Biol. 2 (2003) Article4

The Gene Ontology Consortium: Gene Ontology: Tool for the Unification of Biology, Nature Genet. 25 (2000) 25-29

Dahlquist, K.D., Salomonis, N., Vranizan, K., Lawlor, S.C., Conklin, B.R.: GenMAPP, a New Tool for Viewing and Analyzing Microarray Data on Biological Pathways. Nat. Genet. 31 (2002) 19-20

Ramakrishnan, N., Antoniotti, M., Mishra, B.: Reconstructing Formal Temporal Models of Cellular Events using the GO Process Ontology. Bio-Ontologies SIG Meeting, ISMB 2005 Detroit, U.S.A. (2005)

Chatziioannou, A., Moulos, P., Aidinis, V.: ANDROMEDA: a Pipeline for Versatile Microarray Data Analysis Implemented in MATLAB. (2007) submitted.

Juanita Martinez, M., Aragon, A.D., Rodriguez, A.L., Weber, J.M., Timlin, J.A., Sinclair, M.B., Haaland, D.M., Werner–Washburne, M.: Identification and Removal of Contaminating Fluorescence from Commercial and in–house Printed DNA Microarrays. Nucl. Acids Res. 31 (2003) e18

Scharpf, R.B., Iacobuzio-Donahue, C.A., Sneddon, J.B., Parmigiani, G.: When Should One Subtract Background Fluorescence in Two Color Microarrays? Collection of Biostatistics Research Archive (2005)

Hoffmann, R., Seidl, T., Dugas, M.: Profound Effect of Normalization on Detection of Differentially Expressed Genes in Oligonucleotide Microarray Data Analysis. Genome Biol. 3 (2002) RESEARCH0033

Dudoit, S., Yang, Y.H., Speed, T., Callow, M.J.: Statistical Methods for Identifying Differentially Expressed Genes in Replicated cDNA Microarray Experiments. Statistica Sinica 12 (2002) 111-139

Kerr, M.K., Martin, M., Churchill, G.A.: Analysis of Variance for Gene Expression Microarray Data. J. Computational Biol. 7 (2000), 819-837

Benjamini, Y., Hochberg, Y.: Controlling the False Discovery Rate: a Practical and Powerful Approach to Multiple Testing. J. R. Statist. Soc. 57 (1995) 289-300

Storey, J.D., Tibshirani, R.: Statistical Significance for Genomewide Studies, Proc. Nat. Acad. Sci. 100 (2003) 9440-9445

Hirji, K.F.: Exact Analysis of Discrete Data, Chapman & Hall/CRC (2005)

Subramanian, A., Tamayo, P., Mootha, V.K., Mukherjee, S., Ebert, B.L., Gillette, M.A., Paulovich, A., Pomeroy, S.L., Golub, T.R., Lander, E.S., Mesirov, J.P.: Gene Set Enrichment Analysis: A Knowledge–Based Approach for Interpreting Genome–Wide Expression Profiles. Proc. Nat. Acad. Sci. 102 (2005) 15545-15550

Barry, W.T., Nobel, A.B., Wright, F.A.: Significance Analysis of Functional Categories in Gene Expression Studies: A Structured Permutation Approach. Bioinformatics 21 (2005) 1943-1949

Conover, W.J.: Practical Nonparametric Statistics. Wiley (1980)

Gene Expression Analysis for the Identification of Genes Involved in Early Tumour Development

Stefano Forte[1,3,4], *Salvatore Scarpulla*[1], *Alessandro Laganà*[3,4], *Lorenzo Memeo*[2], and *Massimo Gulisano*[1]

1 Fondazione IOM, Catania, Italy, s.forte@fondazioneiom.it
Home page: http://www.fondazioneiom.it
2 Department of Experimental Oncology, Mediterranean Institute of Oncology, Catania, Italy
3 Departments of Mathematics and Computer Science, University of Catania, Italy
4 Departments of Biomedical Sciences, University of Catania, Italy

Abstract. Prostatic tissues can undergo to cancer insurgence and prostate cancer is one of the most common types of malignancies affecting adult men in the United States. Primary adenocarcinoma of the seminal vesi-cles (SVCA) is a very rare neoplasm with only 48 histologically confirmed cases reported in the European and United States literature. Prostatic tissues, seminal vesicles and epididymis belongs all to the same microenvironment, shows a very close morphology and share the same embryological origin. Despite these common features the rate of cancer occurrence is very different. The understanding of molecular differences between non neoplastic prostatic tissues and non neoplastic epididymis or seminal vesicles may suggest potential mechanisms of resistance to tumour occurrence. The comparison of expression patterns of non neoplastic prostatic and seminal vesicles tissues to identify differentially expressed genes can help researchers in the identification of biological actors involved in the early stages of the tumour development.

1 Introduction

Prostate cancer is one of the most common types of malignancies affecting adult men in the United States [1]. When prostate adenocarcinoma is diagnosed in early stage, it is usually controlled locally by surgery and/or local radiation therapy. However, patients with tumour recurrences and those with advanced prostate cancer, including metastasis to bone and lymph nodes, are limited to hormonal ablation therapy. Unfortunately, there are no effective therapeutic measures once the disease reaches a hormone-refractory state. Primary adenocarcinoma of the seminal vesicles (SVCA) is a very rare neoplasm with only 48 histologically confirmed cases reported in the European and United States literature [2]. The ability to identify SVCA accurately is helpful in clinical management because even in the presence of advanced disease, prompt diagnosis and treatment have been associated with improved long-term survival [2] [3].
The understanding of molecular differences between non neoplastic prostatic tissues and non neoplastic epididymis or seminal vesicles may suggest potential

mechanisms of resistance to tumour occurrence. We decided to compare expression patterns of non neoplastic prostatic and seminal vesicles tissues to identify differentially expressed genes. These tissues have been selected due to their common embryological origin, their similar microenvironment and morphology.

2 Methods

Expression profiles were obtained from the public database Gene Expression Omnibus (GEO) available on the internet on the National Center for Biotechnology Information (NCBI) website. Two Geo dataset were selected: one, GDS 1085, contains expression profiles from 4 different non neoplastic prostatic tissues while the other, GDS 1086, contains expression profiles from one normal epididymis and 3 normal seminal vesicles. All 8 profiles were obtained by the use of the same non-commercial cDNA microarray platform (GPL1823) which features 43008 spots. Total mRNA extracted from the described tissues was hybridized with the same common reference sample. Expression values where log2 ratio transformed and imported into Microsoft Excel as Expression table using BRB-ArrayTools version 3.5.0 import facility [4]. Expression values were normalized using scaling on the median value for each array. Genes were filtered for fold change (1.5 FC in either direction from gene's median value), missing values (a maximum of 50The class comparison between groups of arrays procedures was used for the identification of differentially expressed genes. The used significance threshold of univariate tests was 0.001. The first groups of array were populated by profiles of prostatic tissues while the second group was populated by profiles of epididymis and seminal vesicles tissues. A number of differentially expressed genes were selected using this t-test based analysis and were divided in two groups according to the organ specific overexpression (fig. 1 and table 1)

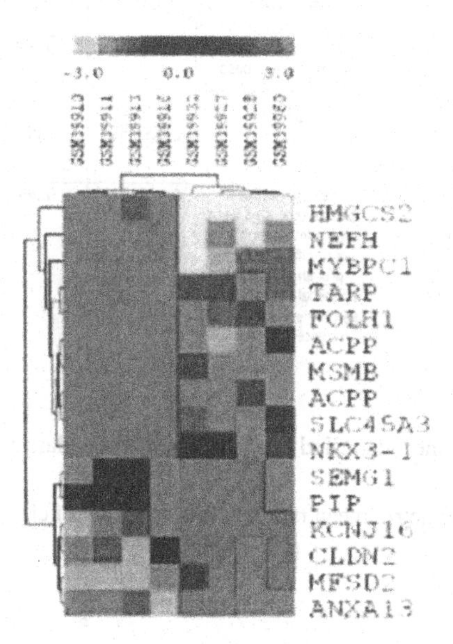

Fig.1. Graphic representation of 16 genes expression levels.

The first 10 genes are over expressed in prostatic tissues, while the last 6 are over expressed in seminal vesicles.

Name	Symbol	SPLocal	p-value	GMR of cl. 1	GMR of cl. 2	Ratio of geom means
Myosin binding protein C, slow type	MYBPC1	membrane; multi-pass membrane protein.	0.0002108	43,562	0.181	240.674
Acid phosphatase, prostate	ACPP		5.8e-06	74,453	0.339	219,625
Acid phosphatase, prostate	ACPP		0.0005532	72,984	0.374	195,144
Microseminoprotein, beta-	MSMB	secreted protein. note=sperm surface.	0.0008328	78,344	0.460	170,313
Neurofilament, heavy polypeptide 200kDa	NEFH		8.69e-05	17,036	0.106	160,717
Solute carrier family 45, member 3	SLC45A3		8.08e-05	60,948	0.474	128,582
TCR gamma alternate reading frame protein	TARP		1,61e-05	49.97	0.397	125,869
3-hydroxy-3-methylglutaryl-Coenzyme A synthase 2 (mitochondrial)	HMGCS2	mitochondrion.	0.0006108	6,836	0.056	122,071
Folate hydrolase (prostate-specific membrane antigen) 1	FOLH1		0.0001034	43,344	0.373	116,204
NK3 transcription factor related, locus 1 (Drosophila)	NKX3-1	nucleus.	3.24e-05	68,534	0.625	109.654
Semenogelin I	SEMG1	secreted protein.	0.0002206	0.788	203.516	0.004
Potassium inwardly-rectifying channel, subfamily J, member 16	KCNJ16	membrane; multi-pass membrane protein.	0.0002225	0.238	49.488	0.005
Prolactin-induced protein	PIP	secreted protein (probable).	0.0001753	0.773	81.950	0.009
Claudin 2	CLDN2	cell junction: tight junction; multi-pass membrane protein (by similarity).	0.0004564	0.367	24.268	0.015
Major facilitator superfamily domain containing 2	MFSD2		0.0002881	0.203	7.245	0.028
Annexin A13	ANXA13	cell membrane. note=associated with the plasma membrane of undifferentiated, proliferating crypt epithelial cells as well as differentiated villus enterocytes.	0.0001496	0.260	8.748	00:03

Table 1. Genes identified using the t-test based class comparison.

GMR of cl. 1 is the Geometric mean ratio of class 1 - GMR of cl. 2 is the Geometric mean ratio of class 2

3 Results

3.1 Genes with increased expression in non neoplastic prostatic tissue when compared with non neoplastic seminal vesicle tissue

MSMB

Beta-microseminoprotein, prostatic secretory protein (PRPS), or PSP94, a 14-kD protein of prostate origin, is an abundant constituent of the human seminal plasma. It is a product of a single gene transcribed in the prostate but not in the testis or in the seminal vesicles. It is synthesized by the epithelial cell of the prostate gland and secreted into the seminal plasma. The primary structure is known from sequence determinations on the purified seminal plasma protein as well as from the cDNA [5]. The 94-amino acid seminal protein is derived from a pre-polypeptide of 114 amino acids. There is a possible N-glycosylation site on the amino-terminal leader peptide of the pre-polypeptide molecule. The protein was considered to be beta-inhibin because of reports of its ability to suppress FSH secretion by the pituitary gland. Later the inhibin activity of this seminal plasma protein was questioned when highly purified preparations were found to be devoid of this activity. Furthermore, it has no structural similarity to the ovarian inhibins. The expression of the encoded protein is found to be decreased in prostate cancer. Two alternatively spliced transcript variants encoding different isoforms are described for this gene. The use of alternate polyadenylation sites has been found for this gene.

MYBPC1

Myosin binding protein C1 (MYBPC1) is critically involved in muscle differentiation and maintenance and has been implicated in the pathogenesis of severe myopathies. Biochemical analyses demonstrated that MyBPC1 is a shorted-lived proteasomal substrate and its degradation is prevented by overexpression of USP25m but not by other USP25 isoforms

HMGCS2

HMGCS2, the gene that regulates ketone body production, is expressed in liver and several extrahepatic tissues, such as the colon. In CaCo-2 colonic epithelial cells, the expression of this gene increases with cell differentiation. Accordingly, immunohistochemistry with specific antibodies shows that HMGCS2 is expressed mainly in differentiated cells of human colonic epithelium. HMGCS2 is a direct target of c-Myc, which represses HMGCS2 transcriptional activity. c-Myc transrepression is mediated by blockade of the transactivating activity of Miz-1, which occurs mainly through a Sp1-binding site in the proximal promoter of the gene. Accordingly, the expression of human HMGCS2 is down-regulated in 90

NKX3.1

Nkx3.1, is a homeobox gene specifically expressed in the prostate epithelium.
NKX3.1 is one of the earliest markers for prostate development and is continuously expressed at all stages during prostate development and in adulthood [6]. Human NKX3.1 maps to chromosome 8p21, a region that frequently undergoes loss of heterozygosity (LOH) at early stages of prostate carcinogenesis [7], [8]. Nkx3.1 mutant mice develop prostatic hyperplasia and dysplasia. However, these early lesions failed to progress to metastatic cancers [6] [9] consistent with a role for Nkx3.1 inactivation in prostate cancer initiation. PTEN loss causes reduced NKX3.1 expression in both murine and human prostate cancers. Restoration of Nkx3.1

expression in vivo in Pten null epithelium leads to decreased cell proliferation, increased cell death, and prevention of tumour initiation [10] [11].

FOLH1

Prostate specific membrane antigen (PSMA), also known as folate hydrolase (FOLH1) is a 100 kDa glycoprotein with elevated expression in prostate epithelial tissue. Expression of PSMA is up regulated as prostate tumour grade increases and is found in the vasculature of many tumours, with no presence in benign tissues

TARP

The T-cell receptor? alternate reading frame protein (TARP) is a 7-kDa mitochondrial product that is specifically expressed by normal and transformed prostatic epithelial cells, which has also been found present in some breast cancer cell lines [12][13]. Recently, two peptide epitopes from TARP were described (TARP4-13 and TARP27-35), which were shown to induce antitumor CTL responses in vitro. Furthermore, circulating CD8+ T lymphocytes from peripheral blood of prostate cancer patients were able to recognize these CTL epitopes, suggesting that immune responses to TARP are vivo.

NEFH and SLC45A3

Neurofilament heavy polypeptide (NEFH), a 200 kDa protein, is involved in the pathogenesis of sporadic motor neuron disease. Solute Carrier family 45 (SLC45A3) is a 59 kDa prostate-specific membrane protein expressed in both neoplastic and non neoplastic prostatic epithelial cells and it is up regulated by androgen

3.2 Genes with increased expression in non neoplastic seminal vesicle tissue when compared with non neoplastic prostatic tissue:

Claudin2

Claudins are adhesion molecules present in tight junctions. Lower expression of claudin 2 was seen in breast and prostatic carcinomas, while hepatocellular and renal carcinomas expressed lower levels of claudins 4 and 5. In contrast to epithelial tumours, lymphomas did not express claudins and most soft tissue tumours and naevocytic lesions are negative or show weaker, mainly cytoplasmic positivity for some claudins.

PIP

The PIP gene, localized in the 7q34 region that contains a number of fragile sites such as FRA 7H and FRA TI, codes for gp17/PIP, a protein secreted by breast apocrine tumours Prolactin-inducible protein (PIP), also known as gross cystic disease fluid protein 15, is a predominant secretory protein in various body fluids, including saliva, milk and seminal plasma. PIP expression was examined in normal prostate tissues and in adenocarcinomas of the prostate. Quantitative real-time RT-PCR revealed low-level presence (6)

SEMG1

The protein encoded by this gene is the predominant protein in semen. The encoded secreted protein is involved in the formation of a gel matrix that encases ejaculated spermatozoa. The prostate-specific antigen (PSA) protease processes this protein into smaller peptides, with each possibly having a separate function.

The proteolysis process breaks down the gel matrix and allows the spermatozoa to move more freely. Two transcript variants encoding different isoforms have been found for this gene.

KCNJ16

Potassium channels are present in most mammalian cells, where they participate in a wide range of physiologic responses. The protein encoded by this gene is an integral membrane protein and inward-rectifier type potassium channel. The encoded protein, which has a greater tendency to allow potassium to flow into a cell rather than out of a cell, can form heterodimers with two other inward rectifier type potassium channels. It may be involved in the regulation of fluid and pH balance. Three transcript variants encoding the same protein have been found for this gene.

MFSD2

Encodes for a cellular component integral to membrane

ANXA13

This gene encodes a member of the annexin family. Members of this calcium dependent phospholipid-binding protein family play a role in the regulation of cellular growth and in signal transduction pathways. The specific function of this gene has not yet been determined; however, it is associated with the plasma membrane of undifferentiated, proliferating endothelial cells and differentiated villous enterocytes. Alternatively spliced transcript variants encoding different isoforms have been identified.

4 Conclusions

The proposed approach could be useful to identify genes that may play critical roles in cancerogenesis. The validity of this methodology is confirmed by the literature findings. In fact MSMB, NKX3.1, FOLH1, TARP are reported to be involved in different steps of embryonic development of the prostate and PIP plays a role in seminal vesicles physiology. The other genes highlighted by the analysis are mostly unknown and their role in cancer insurgence and development still requires further investigations. If these genes will demonstrate to play a role in the carcinogenesis process, the suggested approach could be expanded to different organs and systems and could be a powerful tool to provide new insights into molecular basis of tumours.

References

Jermal, A., Tiwari, C., Murray, T.: Cancer statistics. CA. Cancer J. Clin. 54 (2004) 8_29

Ormsby, H., Haskell, R., Ruthven, E., Mylne, G.: Bilateral primary seminal vesicle carcinoma. Pathology. 28 (1996) 196_200

Benson, R., Clark, W., Farrow, G.: Carcinoma of the seminal vesicle. J. Urol. 132 (1984) 483_5

Simon, R.: Analysis of Gene Expression Data Using BRB-Array Tools. Cancer Informatics 2 (2007) 11_17

Mbikay, M., Nolet, S., Fournier, S.: Molecular cloning and sequence of the cDNA for a 94-amino-acid seminal plasma protein secreted by the human prostate. DNA. 6 (1987) 23_9

Bhatia-Gaur, R., Donjacour, A., Sciavolino, P.: Roles for Nkx3.1 in prostate development and cancer. Genes Dev. 15 (1999) 966_77

Voeller, H., Augustus, M., Madike, V.: Coding region of NKX3.1, a prostate-specific homeobox gene on 8p21, is not mutated in human prostate cancers. Cancer Res. 57 (1997) 4455_9

He, W., Sciavolino, P., Wing, J.: A novel human prostate-specific, androgen- regulated homeobox gene (NKX3.1) that maps to 8p21, a region frequently deleted in prostate cancer. Genomics. 43 (1997) 69_77

Abdulkadir, S., Magee, J., Peters, T.: Conditional loss of Nkx3.1 in adult mice induces prostatic intraepithelial neoplasia. Mol. Cell. Biol. 22 (2002) 1495_503

Lei, Q., Jiao, J., Xin, L.: NKX3.1 stabilizes p53, inhibits AKT activation, and blocks prostate cancer initiation caused by PTEN loss. Cancer Cell 9 (2006) 367_78

Jiang, A., Yu, C., Zhang, P.: p53 overexpression represses androgen-mediated in duction of NKX3.1 in a prostate cancer cell line. Exp. Mol. Med. 38 (2006) 625{33

Oh, S., Terabe, M., Pendleton, C.: Human CTLs to wild-type and enhanced epitopes of a novel prostate and breast tumor-associated protein, TARP, lyse human breast cancer cells. Cancer Res 64 (2004) 2610_8

Carlsson, B., Totterman, T., Essand, M.: Generation of cytotoxic T lympho- cytes specific for the prostate and breast tissue antigen TARP. Prostate. 61 (2004) 161_70

Multi-Knowledge: Collaborative Environments for the Extraction of New Knowledge from Heterogenous Medical Data Sources

Michele Amoretti[1], Diego Ardig`o[2], Franco Mercalli[3]

1 Information Engineering Department, University of Parma, Italy
2 Internal Medicine Department, University of Parma, Italy
3 Centre for Scientific Culture "A. Volta", Como, Italy

Abstract. The general aim of the Multi-Knowledge project is to develop a collaborative environment to allow networks of co-operating medical research centres to create, exchange and manipulate new knowledge from heterogeneous data sources.
The Multi-Knowledge service-oriented architecture will enable workflow design and execution based on novel operating procedures to manage and combine heterogeneous data and makes them easily available for the imputation of study algorithms. In this paper we describe the general framework and the pilot application, providing preliminary results of the combined analysis of microarray and clinical data of 50 patients.

1 Introduction

The Multi-Knowledge project [1], which is funded by the European Commission in the context of the Sixth Framework Programme for Research and Technological Development (Project #027106, thematic area Information Society Technologies), starts from the data processing needs of a network of Medical Research Centres, in Europe and USA, partners in the project and cooperating in researches related to the link between metabolic diseases and cardiovascular risks. These needs are mostly related to the integration of three main sources of information: clinical data, patient-specific genomic and proteomic data (in particular data produced through microarray technology), and demographic data.
In this context the aim of Multi-Knowledge is the development and the validation of a knowledge management environment to allow different groups of researchers, dealing with different sources of data and technological and organisational contexts, to create; exchange and manipulate new knowledge in a seamless way. The ambition is also to create a technological and methodological frame that can easily be extended to include additional sources of data and expertise (bio-medical data, images, environmental data), and can be applied to wider sectors of medical research. This innovative scenario is illustrated in figure 1.

Critical and difficult issues addressed in the project are the management of data that are heterogeneous in nature (continuous and categorical, with different order of magnitude, different degree of precision, etc.), origin (statistical programs, manual introduction from an operator, etc.), and coming from different data environments (from the clinical setting to the molecular biology lab). The Multi-Knowledge service-oriented architecture we are developing will enable workflow design and execution based on novel operating procedures to manage and combine heterogeneous data and make them easily available for the imputation of study algorithms.

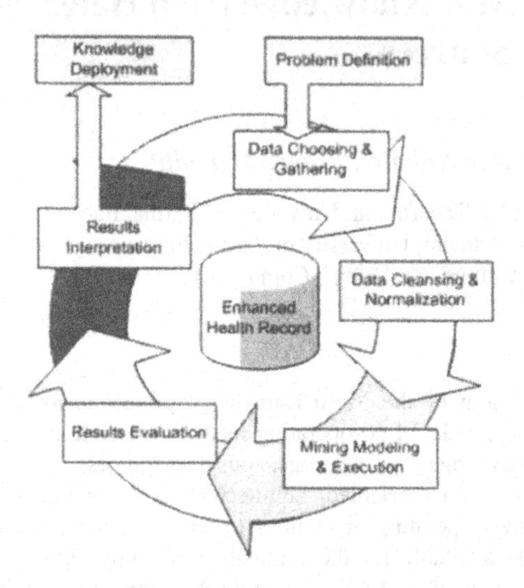

Fig. 1. Multi-Knowledge innovative scenario

The paper is organized as follows. In section 2 we discuss related works in the field of biomedical distributed services. In section 3 we illustrate the general template for Multi-Knowledge workflows, together with some use cases. In section 4 we describe the system prototype and the pilot experiment. In section 5 we illustrate some preliminary results of data analysis experiments related to 50 clinical and microarray data samples.
Finally, we outline some directions for further research and development.

2 Related Works

In the context of clinical services, the European Commission is funding two complementary projects: COCOON [2] and ARTEMIS [3]. COCOON is aimed at setting up a set of regional semantics-based healthcare information infrastructure with the goal of reducing medical errors. ARTEMIS aims to develop a semantic Web Services based interoperability framework for the healthcare domain, building upon a peer-to-peer architecture in order to facilitate the discovery of healthcare Web Services.
The Biological Web Services (BWS) page [4] describes the main services that are available as of March 2006, with appropriate links. Among the services listed by BWS, GeneCruiser [5] is a Web Service for the annotation of microarray data,

developed at the Broad Institute (a research collaboration of MIT, Harvard and its affiliated Hospitals).
GeneCruiser allows users to annotate their genomic data by mapping microarray feature identifiers to gene identifiers from databases, such as UniGene, while providing links to web resources, such as the UCSC Genome Browser. It relies on a regularly updated database that retrieves and indexes the mappings between microarray probes and genomic databases. Genes are identified using the Life Sciences Identifier standard.
A more complex example of Web Service-oriented architecture providing transparent access to biomedical applications on distributed computational resources is the National Biomedical Computation Resource (NBCR) [6], which is based on Grid technologies such as Globus Toolkit. NBCR users are allowed to design and execute complex biomedical analysis pipelines or workflows of services.
Compared to these initiatives, the Multi-Knowledge project is a step forward since its objective is the creation of collaborative environments in which many kinds of actors (physicians, biomedical researchers, etc.) participate in the workflow execution.

3 Multi-KnowledgeWorkflows

Multi-Knowledge experiments represent process instances, and each experiment step represents an activity. Experiment steps are defined by and conducted under the responsibility of a research team, coordinated by a Principal Investigator. Starting from a patients data sample, usually defined and collected in the first work phases, the experiment is set to conduct successive data analysis cycles, aimed at extracting new knowledge through the exploitation of full integration among heterogeneous data clinical, demographical, genomic and proteomic managed by a diverse set of researchers. Biomedical researchers and biostatisticians are the major members of the research team. The data sample is populated through the execution of relevant data collection steps.
As above mentioned, data collection steps are normally the first steps to be conducted. Data analysis steps form the core of the experiments analysis cycle. Through them, the data sample is successively analysed by different classes of researchers having different "scientific cultures" and backgrounds that use different analysis tools, work in different environments, at geographically dispersed sites. Each of the data analysis step may generate new knowledge elements that contribute to create and successively expand an experiment-related body of knowledge (EBoK). Based on an analysis of the EBoK (performed from their different scientific point of views) research team members can propose the execution of additional experiment steps or to further carry on the process.
Thus, the Multi-Knowledge workflow system:
– introduces experiment steps that are conducted by different researchers with diverse scientific background and cultures;
– supports the need of passing control back and forth from different researchers to perform data analysis steps relating to completely different mathematical foundations;

– allows the experiments consist of dynamical cycles of data collection and analysis that aim at progressively achieving the scientific goal initially stated for the experiment;
– concerns the collaboration among different teams, which are independently performing experiments in related areas.
The general template for Multi-Knowledge workflows complies with the activity diagram in figure 2.

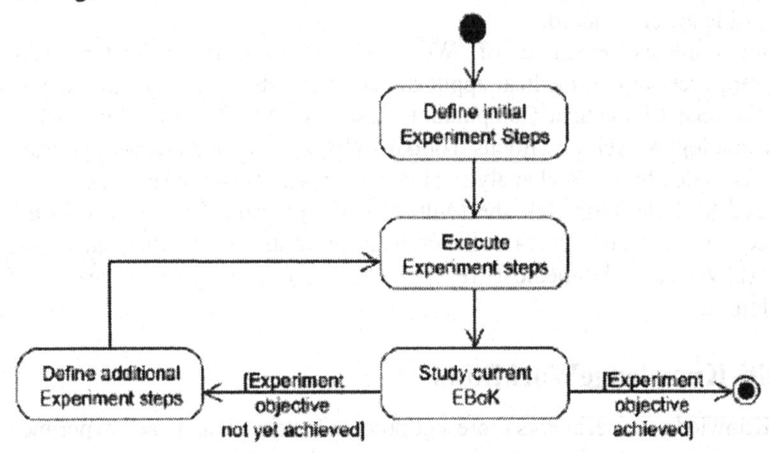

Fig. 2. Multi-Knowledge workflow template

When a team member, possibly after receiving a suggestion sent by another team member or by the principal investigator, decides to execute an experiment step, he/she:
– revises the proposed experiment step definition and possibly improves it based on her/his specific knowledge;
– executes the experiment step;
– adds an annotation, presenting the motivations for the experiment step as well as comments on steps execution and outcome.
The workflow engine reacts by logging the experiment step that has been executed, in terms of task identifier, ask parameters and used data set, and by recording the annotations produced by the team member that executed the Step.
Moreover, team members are allowed to browse the current experiment status, consisting of all the experiment steps conducted so far and related information. The system offers a comprehensive view of this knowledge, allowing to choose and extract graphic representations of different (including intermediate) steps of the experiment, to define and print reports based on the experiment status or on specific parts of it, and to perform statistics on logging information coming from different experiments managed within the system, in order to extract performance measures and identify best practices.

4 Pilot Experiment and System Prototypes

In the first instance of MK pilot study clinical, laboratory, instrumental and genomic information has been collected from 50 subjects by the Department of Internal

Medicine of the University of Parma. The sample has been used to validate the first MK IT system prototype, in particular the system modules related to data collection and normalization, and to define preliminary OLAP/mining models for heterogenous data analysis.

On the server side, the following modules were deployed:

– the **Portal**, which is the point of access to the knowledge extraction system;

– the **Data Collection and Normalization (MK-DCNS)**, whose goal is to provide a common integration service bus and a set of specialized application adapters for the collection, integration and normalization of heterogeneous data from heterogeneous data sources.

Different kinds of biological data can be inserted into the MK-DCNS. Referring to RNA expression arrays and protein arrays, microarray measurements are given as a set of feature extraction (FE) files and an indication of which columns from them to use. Each FE file represents one experiment and contains all the data derived from that microarray. Each expression FE file contains data on about 40000 genes and each protein FE file contains data on about 100 proteins. Furthermore, metabolomics data are given as tab delimited text files with two columns. The first column contains the metabolite description and the second column contains the corresponding numerical values and units. Finally, IMT and FMD data from each patient are entered to the system either manually through a GUI or through a tab delimited text file. If the second option is used, which data to use from within the file will also be supplied. In addition some calculated variables may be specified which will automatically be calculated from the data.

Data Analysis and Visualization modules were deployed on several client machines, distributed worldwide, and used to perform biostatistical analyses on the integrated and normalized data retrieved from the MK-DCNS. A typical analysis task is the following one:

1. the user imports the data;
2. the user partitions the data according to one of the phenotypes, i.e. all people with BMI below XXX are in class A and all people with a BMI above XXX are in class B;
3. with the above partition the user runs a differential expression analysis to see which genes are differentially expressed in the 2 groups;
4. the user runs a GO (Gene Ontology) analysis to see which GO terms are enriched in the above ranked list of genes;
5. results are graphically visualized.

Other types of analysis which can be performed are classification, clustering, class discovery, and sequence motifs finding.

The second instance of the pilot experiment will require further recruitment, up to a estimated sample of about 150-200 subjects. This new study will be performed to test an enhanced MK IT system prototype, including the following modules:

– the Reporting module, providing two web interfaces respectively for the definition of report structures, and for the creation and publication of useful reports after each data analysis experiment step;

– the Workflow Management module, allowing to define and control of the experiments.

All modules have been or are being developed by different IT partners of the Multi-Knowledge project. System integration is facilitated by the adoption of Web Service technologies. One important aspect is security, including user authentication and authorization (with a role-based policy) and sample data anonymization and protection.

5 Preliminary Results

The implemented components of the MKsystem have been used to perform preliminary statistical analyses on the first MK pilot experiment. All the major statistical scenarios have been tested.

As first step, we tried to reproduce results already known from the medical literature, such as the difference in gene expression profile between the two genders. The major differences between males and females in already published studies are related to the expression of genes codified in the sexual chromosomes, being Y-chromosome genes expressed exclusively in men and X-linked genes significantly more expressed in women. Using the MK overabundance analysis algorithm [7], we identified the list of genes differentially expressed between the two genders. As shown in figure 3, the algorithm discriminated almost perfectly (only one misplacement) between the two genders and the differentially expressed genes driving this difference were several genderrelated genes (especially genes from the Y chromosome), as it is evident from the heatmap representation of the 50 most differentially expressed.

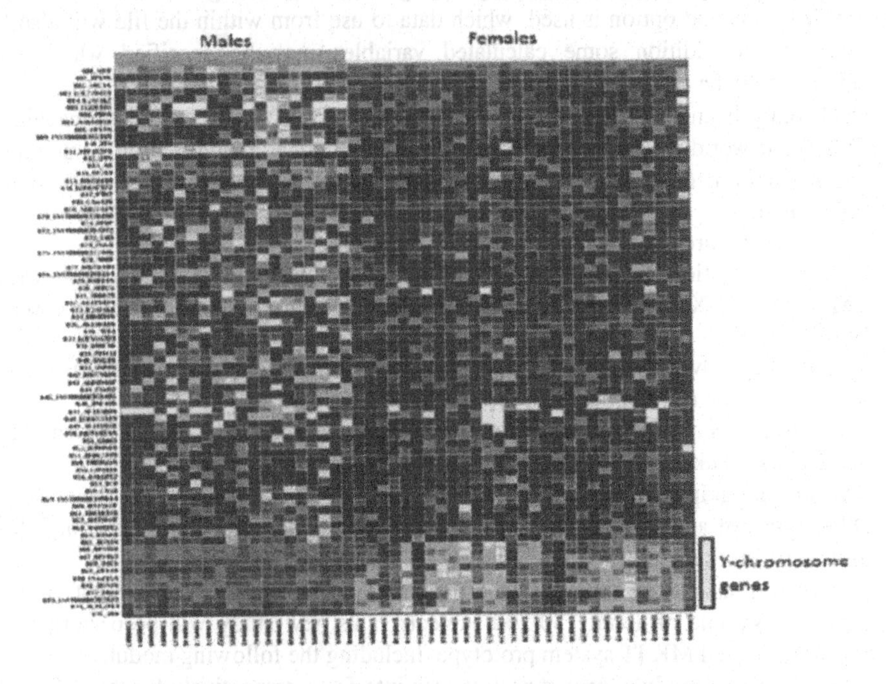

Fig. 3. MK overabundance analysis algorithm results.

As next step, we analyzed the differential expression between two groups of patients selected on the basis of a clinical partition. We chose smoking habit as a clinically relevant variable with a natural dichotomous distribution. We identified two groups of individuals (current smokers and never smokers) and we explored the differential

expression between the two groups using partitioning function and overabundance algorithm implemented in the MK system.
The results of the analysis show a significant difference in expression profile of several genes between smokers and non-smokers, and the subsequent GO analysis identifies GO terms related to inflammatory response (such as "defense response" and "immune response") as the most over-expressed in smokers. Results were also confirmed using external, validated GO analysis tools (like EASE [8]). On the overall population in study we also analyzed quantitative parameters of smoking, such as number of smoked cigarettes per-week (cig/wk). To identify the profile of differentially expressed genes between null to light smokers and heavy smokers, we run a partition search through a novel MK algorithm performing TNoM overabundance analysis [9] in all the possible partitions of the selected variable. The algorithm identified the two cutoff values for cigarettes/week to have the best discrimination between the two groups as less than 35 and more than 126 cig/wk, meaning about less than 5 cigarettes and almost 1 pack per day. Figure 4 illustrates the heatmap representation of the genes most differentially expressed in the two groups showing an extremely good partition between the two groups.

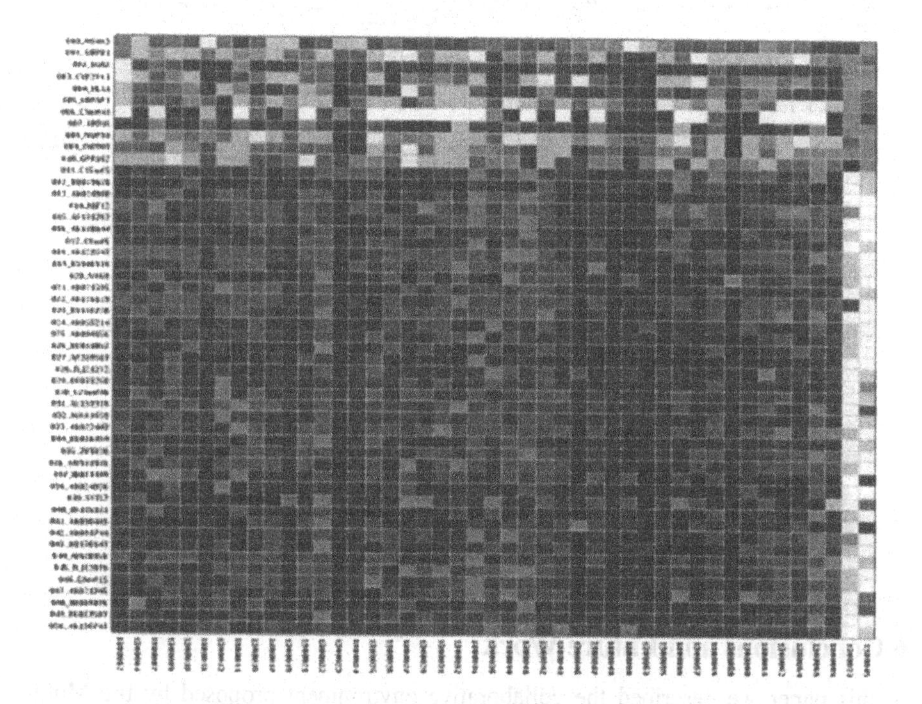

Fig. 4. Difference in expression profile of several genes between smokers and non-smokers.

Significant results in terms of differential expression were also found for several other clinically relevant variables through bi-partition discovery analysis, including plasma LDL cholesterol concentration (the major risk factor for cardiovascular disease), hs-CRP (the major biomarker for atherosclerosis-related inflammation), and IMT (a surrogate markers for early vascular signs of atherosclerosis). In case of IMT

and hs- CRP, the cut-off thresholds identified by the partition algorithm as the best threshold to classify the sample in two groups were 1 mm and 3 mg/L that are well known values used in clinical research to identify subjects with high degree of inflammation and vascular wall damage respectively.
The GO analysis operated on the differentially expressed genes in subjects with high LDL cholesterol compared to low LDL cholesterol showed a significant enrichment of several GO terms related to inflammation and metabolism, as shown in figure 5 using the MK visualization module.

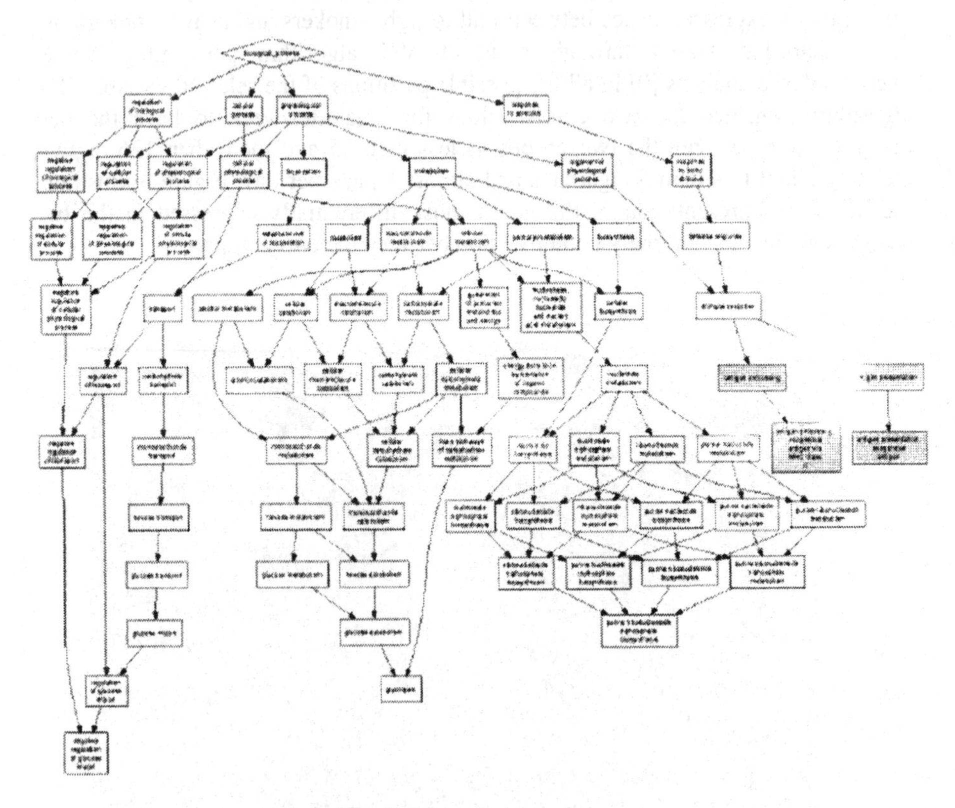

Fig. 5. GO analysis results.

6 Conclusions and Future Work

In this paper we described the collaborative environment proposed by the Multi-Knowledge project. We illustrated the generic template of Multi-Knowledge workflow processes, and the pilot experiment which we are carrying out. We presented the first Multi- Knowledge IT platform prototype, along with preliminary scientific results obtained from the analysis of microarray/clinical data of 50 subjects.

During the second year of the project, theWorkflow Engine module and the Reporting module will be developed and integrated in the prototype. The pilot experiment will be completed considering 100 more samples.

References

Multi-Knowledge Consortium: Multi-Knowledge home page. http://www.multiknowledge.eu
Cocoon consortium: COCOON EU Project homepage. http://www.cocoon-health.com
Artemis Consortium: ARTEMIS EU Project homepage
http://www.srdc.metu.edu.tr/webpage/projects/artemis/index.html
Hull, D.: The Biological Web Services page.
http://taverna.sourceforge.net/index.php?doc=services.html
Liefeld, T. and Reich, M. and Gould, J. and Zhang, P. and Tamayo, P. and Mesirov, J. P.: GeneCruiser: a Web Service for the annotation of microarray data. Bioinformatics **18** (2005) 3681–3682
Krishnan, S. and Baldridge, K. and Greenberg, J. and Stearn, B. and Bhatia, K. An End-to-end Web Services-based Infrastructure for Biomedical Applications. 6th IEEE/ACM International Workshop on Grid Computing, Seattle, Washington, USA, November 2005.
Ben-Dor, A. and Friedman, N. and Yakhini Z. Overabundance Analysis and Class Discovery in Gene Expression Data. Technical Report 2002-50, School of Computer Science & Engineering, Hebrew University , 2002.
EASE: the Expression Analysis Systematic Explorer.
http://david.abcc.ncifcrf.gov/ease/ease.jsp
Ben-Dor, A. and Bruhn, L. and Friedman, N. and Nachman, I. and Schummer, M. and Yakhini,
Z. Tissue classification with gene expression profiles. Journal of Comput. Biol. 2000; 7(3-4):559-83.

Pattern Analysis and Decision Support for Cancer through Clinico-Genomic Profiles

Themis P. Exarchos[1], Nikolaos Giannakeas[1], Yorgos Goletsis[1,2,], Costas Papaloukas[1,3], and Dimitrios I. Fotiadis[1]*

1 Unit of Medical Technology and Intelligent Information Systems, Dept. of Computer Science, University of Ioannina, PO Box 1186, GR 451 10 Ioannina, GREECE
2 Dept. of Economics, University of Ioannina, GR 45110, Ioannina, Greece
3 Dept. of Biological Applications and Technology, University of Ioannina, GR 45110, Ioannina, Greece
*Indicate corresponding author

me01238@cc.uoi.gr, me01310@cc.uoi.gr, goletsis@cc.uoi.gr, papalouk@cc.uoi.gr, fotiadis@cs.uoi.gr

Abstract. Advances in genome technology are playing a growing role in medicine and healthcare. With the development of new technologies and opportunities for large-scale analysis of the genome, genomic data have a clear impact on medicine. Cancer prognostics and therapeutics are among the first major test cases for genomic medicine, given that all types of cancer are related with genomic instability. In this paper we present a novel system for pattern analysis and decision support in cancer. The system integrates clinical data from electronic health records and genomic data. Pattern analysis and data mining methods are applied to these integrated data and the discovered knowledge is used for cancer decision support. Through this integration, conclusions can be drawn for early diagnosis, staging and cancer treatment.

Keywords: Cancer, decision support systems, pattern analysis, profiles, data integration.

1 Introduction

Computer aided medical diagnosis is one of the most important research fields in biomedical engineering. Most of the efforts made, focus on diagnosis based on clinical features. The latest developments in the biomolecular sciences have as a result an explosive growth of biological data available to the scientific community. Bioinformatics provides tools for biological information processing, representing today's key in understanding the molecular basis of physiological and pathological genotypes [1]. The exploitation of bioinformatics for medical diagnosis is an emerging field for the integration of clinical and genomic features and maximizing the information regarding the patient's health status and the quality of the computer aided diagnosis.

Cancer is one of the prominent domains, where this integration is expected to bring significant achievements. As genetic features play significant role in the metabolism and the function of the cells, the integration of genetic information (proteomics-genomics) to cancer related decision support is now perceived by many not as a future trend but rather as a demanding need. The usual patient management in cancer treatment involves several, usually iterative, steps consisting of diagnosis, staging, treatment selection and prognosis. As the patient is usually asked to perform new examinations, diagnosis and staging status can change over time, while treatment selection and prognosis [2] depends on the available findings, response to previous treatment plan and, of course, clinical guidelines. The integration of these evolving and changing data into clinical decision is a hard task which makes fully personalised treatment plan almost impossible. The use of clinical decision support systems (CDSSs) [3] can assist in the processing of the available information and provide accurate staging, personalised treatment selection and prognosis. The development of electronic patient records and of technologies that produce and collect biological information have led to a plethora of data characterizing a specific patient. Although, this might seem beneficial, it can lead to confusion and weakness concerning the data management. The integration of the patient data (quantitative) that are hard to be processed by a human decision maker (the clinician) further imposes the use of CDSSs in personalized medical care [4]. The future vision - but current need - will not include generic treatment plans according to some naive reasoning, but totally personalised treatment based on the clinico-genomic profile of the patient [2].

In this paper we address decision support for cancer, by exploiting clinical data and genomic data. The goal is to perform data integration between medicine and molecular biology, by developing a framework where, clinical and genomic features are appropriately combined in order to handle cancer diseases. The constitution of such a decision support system is based on a) cancer clinical data and b) biological information that is derived from genomic sources. Through this integration, conclusions can be drawn for early diagnosis, staging and effective cancer treatment.

2 Clinical Decision Support using Clinico-genomic Profiles

2.1 Description of the Methodology

Most of the proposed approaches for clinical decision support focus on a single outcome regarding their domain of application. A different approach is to generate profiles associating the input features (e.g. findings) with several outcomes. These profiles include clinical and genomic data along with specific diagnosis, treatment and follow-up recommendations. Profile-based CDSS is based on the fact that patients sharing similar findings are most likely to share the same diagnosis and should have the same treatment and follow-up; the higher this similarity is, the more probable this hypothesis holds. The profiles are created from an initial dataset including several patient cases using a clustering method. Health records of diagnosed and (successfully or unsuccessfully) treated patients, with clear follow-up description, are used to create the profiles. These profiles constitute the core of the CDSSs; each new case that is inserted, is related with one (or more) of these profiles. More specifically, an individual health record containing only findings (and maybe the diagnosis) is matched to the centroids. The matching centroids are examined in order to indicate potential diagnosis (the term diagnosis here refers mainly to the identification of the cancer sub-type). If the diagnosis is confirmed, genetic screening may be proposed to the subject and then, the clusters are further examined, in order to make a decision regarding the preferred treatment and follow-up (Fig. 1).

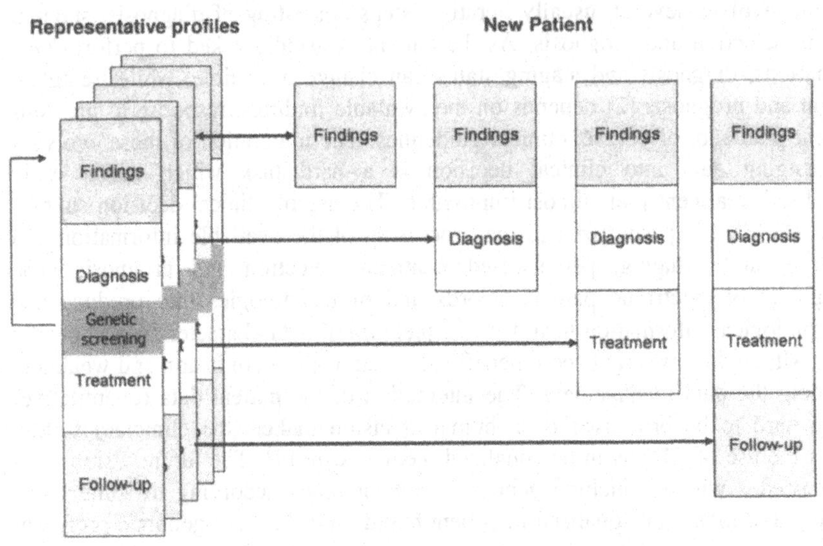

Fig. 1. Decision support based on profiles. Unknown features (diagnosis, treatment, follow-up) of a new case, described only with findings are derived by known features of similar cases.

2.2 General description of the system

Known approaches for the creation of CDSSs are based on the analysis of clinical data using machine learning techniques. This scheme can be expanded to include genomic information, as well. In order to extract a set of profiles, the integration of clinical and genomic data is first required. Then, data analysis is realized in order to discover useful knowledge in the form of profiles. Several techniques and algorithms can be used for data analysis such as neural approaches, statistical analysis, data mining, clustering and others. Data analysis is a two stage procedure: (i) creation of an inference engine (training stage) and (ii) use of this engine for decision support. The type of analysis to be used greatly depends on the available information and the desired outcome. Clustering algorithms can be employed in order to extract patient clinico-genomic profiles. An initial set of records, including clinical and genomic data along with all diagnosis/treatment/follow-up information, must be available for the creation of the inference engine. The records are used for clustering and the centroids of the generated clusters constitute the profiles. These profiles are then used for decision support; new patients with similar clinical and genomic data are assigned to the same cluster, i.e. they share the same profile. Thus, a probable diagnosis, treatment and follow-up, is selected. Both, the creation of the inference engine and the decision support procedure are presented in Fig. 2.

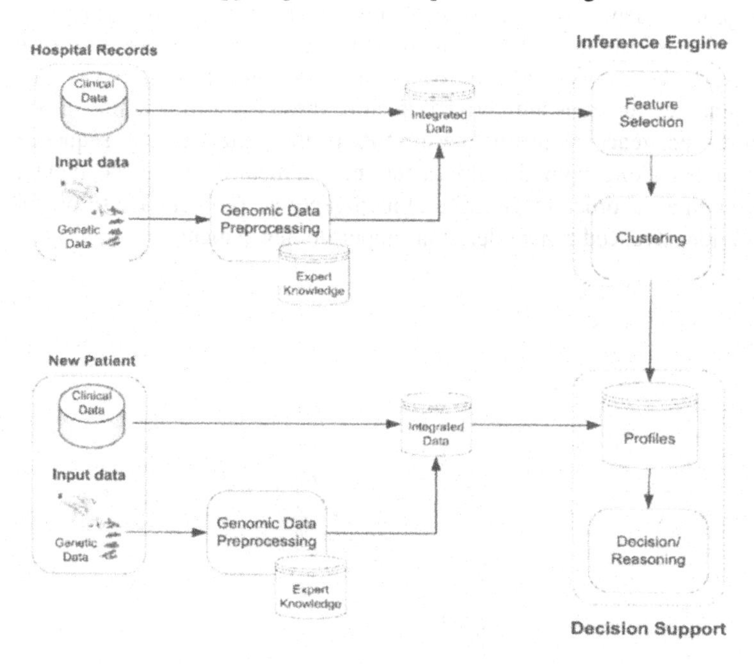

Fig. 2. Representation of a general scheme for the creation of an inference engine of a profile based CDSS, integrating clinical and genomic information. A new case (patient) is applied to the profiles for decision support and reasoning.

2.2.1 Types of data

Clinical data such as demographic details, medical history and laboratory data are usually presented in a structured format, making their analysis an easy task. The proposed method combines the clinical data with biological data, and more specifically, with gene sequence data. An efficient way to process the above gene sequences is to detect Single Nucleotide Polymorphisms (SNPs) [5]. SNPs data are qualitative data providing information about the genomic at a specific locus of a gene. An SNP is a point mutation present in at least 1 % of a population. A point mutation is a substitution of one base pair or a deletion, which means, the respective base pair is missing, or an addition of one base pair. Though several different sequence variants may occur at each considered locus usually one specific variant of the most common sequence is found, an exchange from adenine (A) to guanine (G), for instance. Thus, information is basically given in form of categories denoting the combinations of base pairs for the two chromosomes, e.g. A/G, if the most frequent variant is adenine and the single nucleotide polymorphism is an exchange from adenine to guanine.

2.2.2 Data processing

Since the gene sequence data are not structured, appropriate preprocessing is needed in order to transform them into a more structured format. Gene sequences are acquired from the subjects and based on the SNP information regarding every acquired gene new features are derived. Each of these features contains information regarding the existence or not of these SNPs in the patient's gene sequence. The derived features along with the aforementioned clinical data are the input to the inference engine, in order to generate clinico-genomic profiles. These profiles are able to provide advanced cancer decision support to new patients.

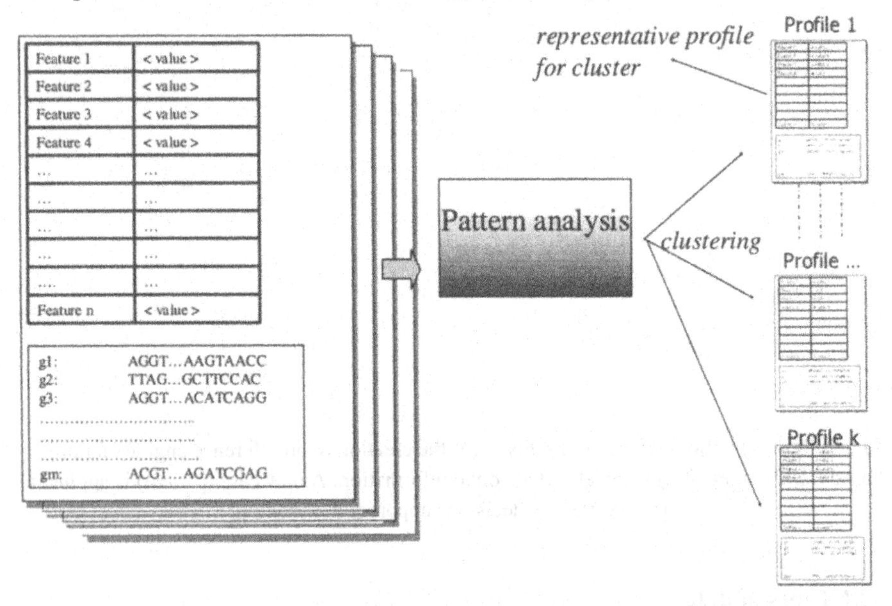

Fig. 3. Schematic representation of clustering integrated clinical and genomic data. The outputs are the representative clinico-genomic profiles for each generated cluster.

The initial dataset (clinical or genomic) is defined by the experts and includes all features that according to their opinion are highly related with the domain at hand (clinical disease). After acquiring the integrated data, a feature selection technique is applied in order to reduce the number of features and remove irrelevant or redundant ones. Since the proposed scheme for decision support focuses on several outcomes, a supervised feature selection technique can not be employed. For this reason, a method based on principal component analysis is used to reduce the number of features [6]. Finally, the reduced set of features is used by a clustering algorithm. k-means algorithm [7] is a promising approach for clustering and can be involved for profile extraction. k-means handles both continuous and discrete data and has low time and space complexity. Also, it provides straightforward distance computation, using the Euclidean distance for continuous data and city-block distance for the discrete data. Furthermore, k-means is an order independent algorithm, since for a given initial distribution of clusters generates the same partition of the data at the end of the partitioning process, irrespective of the order in which the samples are presented to the algorithm. A deficiency of the k-means algorithm is that the number of clusters (profiles) must be predefined, which is not always feasible. Thus, in order to fully automate the profile extraction process, a meta-analysis technique is employed, which automatically calculates the optimal number of profiles [8]. This technique divides the data into 10 sets and performs clustering in each of them. Initially, k is set to 2 and the mean value of the sum of squared errors over the 10 sets is computed. k is increased until the mean value of the sum of squared errors is stabilized or is higher than the previous value of k (k-1). A schematic representation of clustering integrated clinical and genomic features and extracting a set of profiles is shown in Fig. 3.

3 Conclusions and Future work

Several challenges remain, regarding clinical and genomic data integration to facilitate clinical decision support. The opportunities of combining these two types of data are obvious, as they allow obtaining new insights concerning diagnosis, prognosis and treatment. A limitation of this combination is that although data exist, usually their enormous volume and their heterogeneity constitute their analysis and association a very difficult task. The lack of terminological and ontological compatibility, which could be solved by means of a uniform representation is another future challenge. Besides new data models, ontologies are/have to be developed in order to link genomic and clinical data and standards are required to ensure interoperability between disparate data sources.

Acknowledgments. This research is partly funded by the European Commission as part of the project MATCH (IST-2005-027266).

References

Campbell, A.M., Heyer, L.J.: Discovering Genomics, Proteomics and Bioinformatics. Benjamin Cummings, CA, USA (2007).

Goletsis, Y., Exarchos, T.P., Giannakeas, N., Tsipouras, M.G., Fotiadis, D.I.: Integration of clinical and genomic data for decision support in cancer, in Wickramasinghe N., Geisler E. (eds), Encyclopedia of Healthcare Information Systems, Idea Group Publishing, USA, (to be published).

Fotiadis, D.I., Goletsis, Y., Likas, A., & Papadopoulos, A.: Clinical Decision Support Systems. Encyclopedia of Biomedical Engineering, Wiley (2003).

Louie, B., Mork, P., Martin-Sanchez, F., Halevy, A., & Tarczy-Hornoch, P.: Data integration and genomic medicine. Journal of Biomedical Informatics 40 (2007) 5–16.

Sielinski, S. (2005). Similarity measures for clustering SNP and epidemiological data. Technical report of university of Dortmund.

Webb, A.: Statistical Pattern Recognition. Arnold, New York, USA (1999).

Tan, P.N., Steinbach, M., & Kumar, V.: Introduction to Data Mining, Addison Wesley. USA (2005).

Witten, I.H., and Frank, E.: Data Mining: Practical machine learning tools and techniques with java implementations. Morgan Kaufmann, CA, USA (2005).

Section 3

International Workshop on

Rough Sets and Data Mining

Categorization of Musical Instrument Sounds Based on Numerical Parameters

Rory A. Lewis[1] *and Alicja Wieczorkowska*[2]

1 University of North Carolina, 9201 University City Blvd. Charlotte, NC 28223, USA
2 Polish-Japanese Institute of Information Technology, Koszykowa 86, 02-008 Warsaw, Poland

Abstract. In this paper we present methodology of categorization of musical instruments sounds, aiming at the continuing goal of codifying the classiffication of these sounds for automating indexing and retrieval purposes. The proposed categorization is based on numerical parameters. The motivation for this paper is based upon the fallibility of Hornbostel and Sachs generic classiffication scheme, most commonly used for categorization of musical instruments. In eliminating the discrepancies of Hornbostel and Sachs' classiffication of musical sounds we present a procedure that draws categorization from numerical attributes, describing both time domain and spectrum of sound, rather than using classiffication based directly on Hornbostel and Sachs scheme. As a result we propose a categorization system based upon the empirical musical parameters and then incorporating the resultant structure for classiffication rules.

1 Introduction

Categorization of musical instruments into groups and families, although already elaborated in a few ways, is still disputable. Basic categorization, commonly used, is called Sachs and Hornbostel system [7], [14]. It is based on sonorous material producing sound in each instrument. This system was adopted by the Library of Congress [1] and the German Schlagwortnormdatei Decimal Classiffication. They both use the Dewey classiffication system [4, 12]; in 1914 Hornbostel and Sachs devised a classiffication system, based on the Dewey decimal classiffication which essentially classiffied all instruments into strings, wind and percussion, and later further into four categories:

1. Idiophones, where sound is produced by vibration of the body of the instrument
2. Membranophones, where sound produced by the vibration of a membrane
3. Chordophones, where sound is produced by the vibration of strings
4. Aerophones, where sound is produced by vibrating air.

However, in many cases the sound is produced by vibration of various sonorous bodies, for example strings, solid body, i.e. body of the instrument (which can work as resonator amplifying some frequencies) and air contained in the body. Moreover, this classiffication does not strictly reflect a natural division of musical sounds into sustainable and non-sustainable, i.e. containing steady state of the sound or not. For example, percussive instrument produce non-sustainable sounds, but also plucked string produces such a sound, too. Therefore, in our opinion, categorization of musical instrument and their sounds should take articulation into account, i.e. the way how the sound is performed by a player. Our goal was to elaborate such a categorization, based on numerical sound description (i.e. sound parameters that can be automatically calculated for a sound), which produces a clear classiffication of musical instrument sounds, leaving no space for doubts.

2 Instrument Classiffication Based on Numerical Sound Attributes

In order to perform musical instrument sound categorization based on numerical sound parameters, we decided to use a few conditional attributes (LogAt- tack, Harmonicity, Sustainability, SpectralCentroid and TemporalCentroid) and 2 decision attributes: instrument, and articulation. The attributes are based on MPEG-7 low-level sound description [8]. For purposes of music information retrieval, Hornbostel-Sachs is incompatible for a knowledge discovery discourse since it contains exceptions, since it follows a humanistic convention. For example, a piano emits sound when the hammer strikes strings hence, strictly speaking, piano is a percussive instrument. Hornbostel-Sachs categorizes piano as a chordophone, i.e. string instrument. We focus on ffive properties of sound waves that can be calculated for any sound and can di®erentiate. They are: LogAttack, Harmonicity, Sustainability, SpectralCentroid and TemporalCentroid. The ffirst two properties are part of the set of descriptors for audio content description provided in the MPEG-7 standard and have aided us in musical instrument timbre description, audio signature and sound description [17]. The remaining three attributes are based on temporal observations of sound envelopes for singular sound of various instruments and for various playing method.

2.1 LogAttackTime (LAT)

Segments containing short LAT periods cut generic sustained and non-sustained sounds into two separate groups [16, 8]. LAT is the logarithm of the time duration between the points where the signal starts to the point it reaches its stable part [18]. Range is the $log_{10}(1/_{samplingrate})$ and is determined by the length of the signal.

$$LAT = Log_{10}(T1 - T0); \qquad (1)$$

Where $T0$ is the time the signal starts; and $T1$ is the time to reach its sustained part (harmonic space) or maximum part (percussive space).

2.2 AudioHarmonicityType (HRM)

AudioHarmonicityType describes the degree of harmonicity of an audio signal [8]. It includes the weighted conffidence measure, SeriesOfScalarType that handles

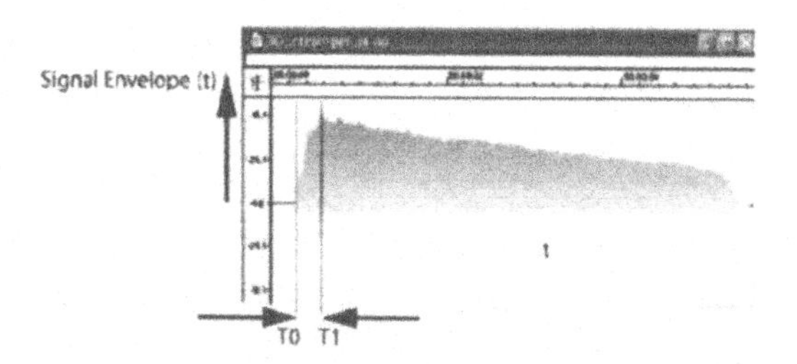

Fig. 1. Illustration of log-attack time.

T0 can be estimated as the time the signal envelope exceeds .02 of its maximum value. *T1 can* be estimated, simply, as the time the signal envelope reaches its maximum value.
portions of signal that lack clear periodicity. AudioHarmonicity combines the ratio of harmonic power to total power: HarmonicRatio, and the frequency of the inharmonic spectrum: UpperLimitOfHarmonicity.
First: We make the Harmonic Ratio *H(i)* the maximum *r(i,k)* in each frame, *i* where a deffinitive periodic signal for *H(i)* =1 and conversely white noise = 0.

$$H(i) = \max r(i; k) \qquad (2)$$

where *r(i,k)* is the normalized cross correlation of frame *i* with lag *k*:

$$r(i,k) = \sum_{j=m}^{m+n+1} s(j)s(j-k) \Bigg/ \left(\sum_{j=m}^{m+n+1} s(j)^2 * \sum_{j=m}^{m+n+1} s(j-k)^2 \right)^{\frac{1}{2}} \qquad (3)$$

where s is the audio signal, $m=i*n$, where i=0, M ; 1=frame index and M = the number of frames, $n=t*sr$, where t = window size (10ms) and sr = sampling rate, $k=1$, $K=lag$, where K=w*sr, w = maximum fundamental period expected (40ms)
Second: Upon obtaining the i) DFTs of $s(j)$ and comb-ffiltered signals $c(j)$ in the AudioSpectrumEnvelope and ii) the power spectra $p(f)$ and $p^0(f)$ in the AudioSpectrumCentroid we take the ratio f_{lim} and calculate the sum of power beyond the frequency for both $s(j)$ and $c(j)$:

$$a(f_{\lim}) = \sum_{f=f_{\lim}}^{f_{\max}} p'(f) \Bigg/ \sum_{f=f_{\lim}}^{f_{\max}} p(f) \qquad (4)$$

where f_{max} is the maximum frequency of the DFT.
Third: Starting where $f_{lim} = f_{max}$ we move down in frequency and stop where the greatest frequency, f_{ulim}'s ratio is smaller than 0.5 and convert it to an octave scale based on 1 kHz:

$$UpperLimitOfHarmonicity = \log 2(f_{ulim}/1000) \qquad (5)$$

2.3 Sustainability (S)

We deffine sustainability into 5 categories based on the degree of dampening or sustainability the instrument can maintain over a maximum period of 7 seconds.

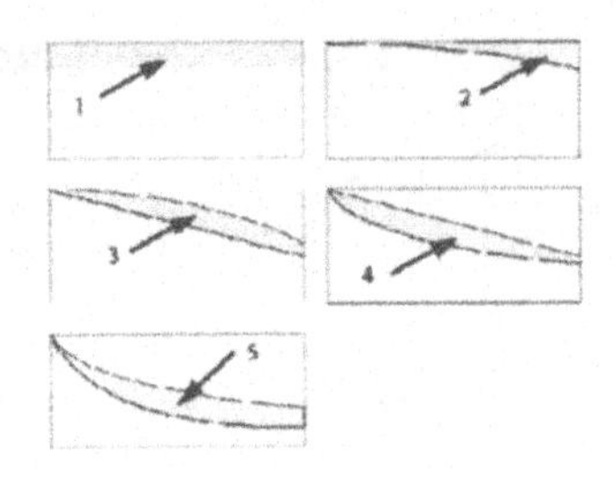

Fig. 2. Five levels of sustainability to severe dampening.

2.4 TemporalCentroid

The TemporalCentroid Descriptor also characterizes the signal envelope, representing where in time the energy of a signal is focused. This Descriptor may, for example, distinguish between a decaying piano note and a sustained organ note, when the lengths and the attacks of the two notes are identical. It is deffined as the time averaged over the energy envelope:

$$TC = \sum_{n=1}^{length(SEnv)} n / sr.SEnv(n) \Bigg/ \sum_{n=1}^{length(SEnv)} SEnv(n) \qquad (6)$$

where *SEnv* is the is the Signal Envelope and *sr* is the Sampling Rate

2.5 SpectralCentroid

The SpectralCentroid Descriptor measures the average frequency, weighted by amplitude, of a spectrum. In cognition applications, it is usually averaged over time:

$$c = \sum c_i \big/ i \Rightarrow ...or... \Rightarrow c_i = \sum f_i a_i \big/ \sum a_i \qquad (7)$$

where c_i is the centroid for one spectral frame, and i is the number of frames for the sound. A spectral frame is some number of samples which is equal to the size of the FFT where each individual centroid of a spectral frame is deffined as the average frequency weighted by amplitudes, divided by the sum of the amplitudes.

3 Experiments

The data set used in our experiments contains 356 tuples, representing sounds of various musical instruments, played with various articulation. The initial data set (see http://www.mir.uncc.edu) contained parameterized sounds for every instrument available on MUMS CDs [11]. The resulting database contains numerical descriptors for 6,300 segmented sounds representing broad range of musical instruments, including orchestral ones, piano, jazz instruments, organ, etc. The MUMS CD's are widely used in musical instrument sound research [2], [19], [20], [21],[22], [23] so they can be considered as a standard. Only a part of these data were used, namely, single sounds for every instrument/articulation, or one from each octave. The risk is that we may have numerous classes with very few representations.

We decided to use decision trees available via Bratko's Orange software that implements the C4.5 with scripting in Python [13], [3]. We run orange/c4.5 with decision attribute which we called sachs-hornbostel-level-1, including 4 classes: aerophones, idiophones, chordophones, and membranophones. Also, we run the classiffier for articulation. We are interested in all objects, which are in the wrong leaves, i.e. misclassiffied objects. Also, apart from c4.5 algorithm, we decided to use a Baysian classiffier.

4 Testing

Observing the database, it is evident in Figure 3that the ffive descriptors by default separate sound into viable clusters. This becomes quite evident when looking at three attributes: LogAttack, Sustainability and SpectralCentroid in relation to the articulation.

Additionally, Figure 3 which is a Polyviz plot which combines RadViz and barchart techniques. It illustrates 1) the clustering of the data points in the middle of the polygon and 2) the distribution along the di®erent dimensions. This becomes quite evident when looking at two attributes: Sustainability and TemporalCentroid in relation to the articulation.

To induce the classiffication rules in the form of decision trees from a set of given examples we used Quinlan's C4.5 algorithm and Oranges implementation of naive Bayesian modeling to compare the results.

4.1 c4.5

Quinlan's C4.5 algorithm [13] constructs a decision tree to form production rules from an unpruned tree. Next a decision tree interpreter classiffies items which the produces the rules. This produced a 10-level tree with 36 leaves. Of the 36 leaves 17 were 100%. The remaining 19 leaves averaged 69.8% with percussive and string instruments imitating each other's attribute the most.

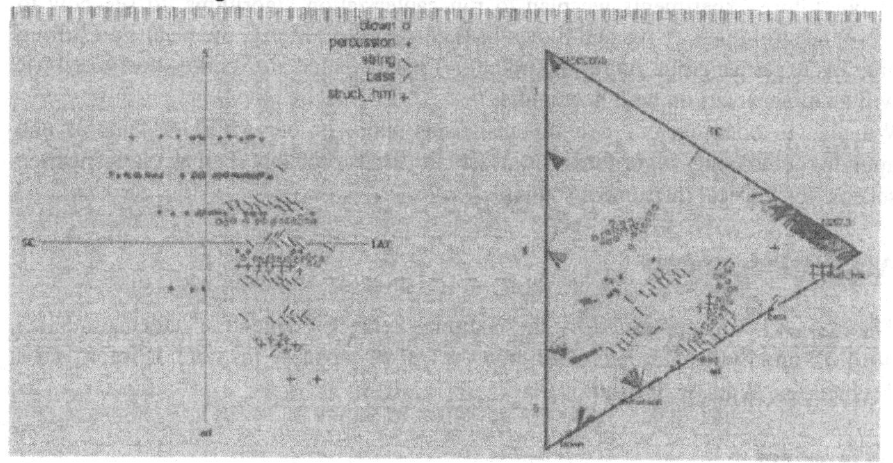

Fig. 3. Illustrations of Linear Projection and Polyviz Clustering.

Figure 3 is a 4 dimensional linear projection to illustrate clustering: LogAttack,

Sustainability, SpectralCentroid and Articulation, Figure 4 illustrate clustering: Sustainability, TemporalCentroid and articulation

4.2 Naive Bayes

Bratco's Naive Bayes based classiffier requires a small amount of training data to estimate the parameters necessary for classiffication. It calculates the probability between input and predictable columns and *naively* assumes that the columns are independent and, by making this assumption, it ignores possible dependencies. This produced a 10-level tree with 61 leaves. Of the 61 leaves 44 were 100%. The remaining 17 leaves averaged 83.2% again with percussive and string instruments imitating each other's attribute the most.

4.3 Rough Sets

Using RSES we discretized and generated rules by using rough set based classiffiers and the LEM2 algorithm. Here we dad results comprising 124 rules but paired it down to 33 after setting parameters of minimum of 90% conffidence with support of no less than 6. Comparing this to the c4.5 17 leaves 44 at 100% with 17 leaves averaging 83.2%, it happens that both c4.5 and rough sets convey the same rules when operating at 90% conffidence with support of no less than 6.

5 Summary and Conclusion

The experiments have proven that implementing temporal attributes that focus on the machine-level view of a signal, and interacting it with MPEG-7 descriptors has proven to yield remarkably strong results. However, keeping this in mind, it has also revealed music information's most problematic di®erentiation of instruments, ffirst the string/percussive cluster and second the blown/bass clusters.
In our future experiments we plan to run cauterization algorithms, to see how to solve the string/percussive and blown/bass clusters. In essence, the results are strong and this is pleasing but, more importantly, it has revealed an Achilles heal that MIR will no doubt focus on with new intent.
We plan to continue our experiments, using more of our MPEG-7 features and applying clustering algorithms in order to ffind probably better classiffication scheme for musical instrument sounds.

Acknowledgements

This research was supported by the National Science Foundation under grant IIS-0414815 and by the Research Center at the Polish-Japanese Institute of Information Technology, Warsaw, Poland.

References

Brenne, M.: Storage and retrieval of musical documents in a FRBR-based library catalogue: Thesis, Oslo University College Faculty of journalism, library and information science,(2004).
Cosi, P., De Poli, G., and Lauzzana, G. (1994). Auditory Modelling and Self- Organizing Neural Networks for Timbre Classiffication *Journal of New Music Research*, 23, 71{98.

Demsar, J., Zupan, B. and Gregor Leban: http://www.ailab.si/orange

Doerr, M.: Semantic Problems of Thesaurus Mapping: *Journal of Digital Information*. Volume 1, issue 8, Article No. 52, 2001-03-26, 2001{03, (2001).

Eronen, A. and Klapuri, A. (2000) Musical Instrument Recognition Using Cepstral Coefficients and Temporal Features. *Proceedings of the IEEE Inter- national Conference on Acoustics, Speech and Signal Processing ICASSP 2000* (753{756). Plymouth, MA.

Fujinaga, I. and McMillan, K. (2000). Realtime recognition of orchestral instruments. *Proceedings of the International Computer Music Conference* (141{143).

Hornbostel, E. M. V., Sachs, C. (1914). Systematik der Musikinstru- mente. Ein Versuch. *Zeitschrift fur Ethnologie*, Vol. 46, No. 4-5, 1914, 553-90, available at http://www.uni-bamberg.de/ppp/ethnomusikologie/HS- Systematik/HS-Systematik

Information Technology Multimedia Content Description Interface Part 4: Audio. ISO/IEC JTC 1/SC 29, Date: 2001-06-9. ISO/IEC FDIS 15938-4:2001(E) ISO/IEC J/TC 1/SC 29/WG 11 Secretariat: ANSI, (2001)

Kaminskyj, I. (2000). Multi-feature Musical Instrument Classiffier. *MikroPoly- phonie* 6 (online journal at http://farben.latrobe.edu.au/).

Martin, K. D. and Kim, Y. E. (1998). 2pMU9. Musical instrument identiffication: A pattern-recognition approach. 136-th meeting of the Acoustical Soc. of America, Norfolk, VA

Opolko, F. and Wapnick, J. (1987). MUMS { McGill University Master Samples. CD's.

Patel, M. and Koch, T. and Doerr, M. and Tsinaraki, C.: Semantic Inter- operability in Digital Library Systems. IST-2002-2.3.1.12 *Technology-enhanced Learning and Access to Cultural Heritage*. UKOLN, University of Bath, (2005).

Quinlan, J.R. 2pMU9. Bagging, boosting, and C4. 5. Proceedings of the Thirteenth National Conference on Artifficial Intelligence. Volume 725, 730, (1996).

SIL International (1999). LinguaLinks Library. Version 3.5. Published on CD-ROM, 1999. TheInternet: http://www.silinternational.org/LinguaLinks/Anthropology/ExpnddEthnmsclgyCtgrCltrlMtrls/MusicalInstrumentsSubcategorie.htm

Wieczorkowska, A. (1999). Rough Sets as a Tool for Audio Signal Classiffication. In Z. W. Ras, A. Skowron (Eds.), *Foundations of Intelligent Systems* (pp. 367{ 375). LNCS/LNAI 1609, Springer.

Gomez, E. and Gouyon, F. and Herrera, P. and Amatriain, X.: Using and enhancing the current MPEG-7 standard for a music content processing tool, *Proceedings of the 114th Audio Engineering Society Convention*, Amsterdam, The Netherlands, March, (2003).

Wieczorkowska, A., Wr¶oblewski, J., Synak, P. and Sl»ezak, D.: Application of temporal descriptors to musical instrument sound recognition: in *Proceedings of the International Computer Music Conference* (ICMC'00), Berlin, Germany, (2004).

Peeters, G., McAdams, S. and Herrera, P.: Instrument sound description in the context of MPEG-7: in *Proceedings of the International Computer Music Conference* (ICMC'00), Berlin, Germany, (2000).

Eronen, A. and Klapuri, A. (2000) Musical Instrument Recognition Using Cep- stral Coe±cients and Temporal Features. Proceedings of the IEEE International Conference on Acoustics, Speech and Signal Processing ICASSP 2000 (753{756). Plymouth, MA.

Fujinaga, I. and McMillan, K. (2000). Realtime recognition of orchestral instruments. Proceedings of the International Computer Music Conference (141{ 143).

Kaminskyj, I. (2000). Multi-feature Musical Instrument Classiffier. *MikroPoly- phonie* 6 (online journal at http://farben.latrobe.edu.au/).

Martin, K. D. and Kim, Y. E. (1998). 2pMU9. Musical instrument identiffica- tion: A pattern-recognition approach.136-th meeting of the Acoustical Soc. of America, Norfolk, VA.

Wieczorkowska, A. (1999). Rough Sets as a Tool for Audio Signal Classiffication. In Z. W. Ras, A. Skowron (Eds.), *Foundations of Intelligent Systems* (pp. 367{ 375). LNCS/LNAI 1609, Springer.

Evaluating the Utility of Web-Based Consumer Support Tools Using Rough Sets

Timothy Maciag[1], Daryl H. Hepting[1], Dominik Slezak[2], and Robert J. Hilderman[1]

1 University of Regina, Department of Computer Science, maciagt@cs.uregina.ca, hepting@cs.uregina.ca, robert.hilderman@uregina.ca
2 Infobright Inc., slezak@infobright.com

Abstract. On the Web, many popular e-commerce sites provide consumers with decision support tools to assist them in their commerce-related decision-making. Many consumers will rank the utility of these tools quite highly. Data obtained from web usage mining analyses, which may provide knowledge about a user's online experiences, could help indicate the utility of these tools. This type of analysis could provide insight into whether provided tools are adequately assisting consumers in conducting their online shopping activities or if new or additional enhancements need consideration. Although some research in this regard has been described in previous literature, there is still much that can be done. The authors of this paper hypothesize that a measurement of consumer *decision accuracy*, i.e. a measurement indicating the success in which consumers are able to find items matching specified preferences, could help indicate the utility of these tools. This paper describes a procedure developed towards this goal using elements of rough set theory. The authors evaluated the procedure using two support tools, one based on a tool developed by the US-EPA and the other developed by one of the authors called *cogito*. Results from the evaluation did provide interesting insights on the utility of both support tools. Although it was shown that the *cogito* tool obtained slightly higher decision accuracy, both tools could be improved from additional enhancements. Details of the procedure developed and results obtained from the evaluation will be provided. Opportunities for future work are also discussed.

1 Introduction

Each day millions of consumers connect to the World Wide Web to conduct a variety of e-commerce activities. As such, many current and popular e-commerce sites attempt to provide their consumers with enhanced support tools that assist them in these activities. As many of these support tools evolve, competition to provide consumers with the best possible support tools, thus the best possible shopping experience, intensifies (Ha 2002). However, the question remains (Spiliopoulou 2000):

How do we evaluate the utility of these support tools? Techniques in web data mining, which provide the theoretical foundations for discovery of interesting patterns on the web, could assist in this task.
Evaluating the utility of these support tools may provide online retailers with an indication on the success at which consumers are able to conduct their shopping activities while using the tools provided to them. The procedure described in this paper, which uses concepts of web data mining and elements of rough sets, will provide online retailers the ability to evaluate the utility of their tools by enabling them to conduct comparisons and evaluations with ones provided by their competitors.

1.1 Web Data Mining

Web data mining, or simply web mining, is an area of research that deals with finding interesting patterns in online data (Kosala and Blockeel 2000). Web mining research has traditionally been split into three taxonomies: web content mining, web structure mining, and web usage mining. Here, web usage mining, which refers to the process of analyzing user usage patterns online (Srivastava et al. 2000), is of specific interest.
Web usage mining research has received significant interest as of late (Pierrakos et al. 2003). The data obtained from web usage mining analyses, including data obtained from user logs and registration forms (Zhou et al. 2004), could aid online retailers form knowledge about the utility of the support tools they provide (Spiliopoulou 2000). Knowledge obtained from this type of analysis could help indicate if improvements in the design are required, e.g. total revision of the design or incorporation of additional enhancements, such as personalization (Erininaki and Vazirgiannis 2003; Holland et al. 2003; Maciag et al. 2007). Although some work has been done in this regard (Spiliopoulou 2000; Pu and Chen 2005), there is still much research to be done. One method to evaluate the utility of online support tools could be to conduct an analysis of user *decision accuracy* (Pu and Chen 2005).

1.2 Decision Accuracy

Pu and Chen (2005) describe the concept of decision accuracy as a way to evaluate the success at which consumers are able to find items that match specified preferences. Consumers are said to have higher decision accuracy if the distance between their actual and preferred item selections are minimal to none. The authors of this paper hypothesize that higher decision accuracies may indicate more satisfying consumer shopping experiences, thus realizing a higher utility of the support tools used.
In this paper, the authors describe a procedure to discover a measurement of utility by analyzing user decision accuracy using techniques in rough sets. Rough sets (Pawlak, 1991) have been researched and used in a variety of data mining applications over the last several years (Goebel and Gruenwald 1999; Beaubouef et al. 2004; Lingras and West 2004; Slezak 2005; Chimphlee et al. 2006). In the context of this paper, rough set reduction and classification techniques are used. Obtained measurements of accuracy and coverage are analyzed to indicate percentages of user decision accuracy.

2 Background

Previously, the authors conducted an evaluation of support tools for environmentally preferable purchasing and have analyzed results obtained (Maciag 2005). Two support tools were used in the evaluation, one based on a support tool provided by the
United States Environmental Protection Agency (US-EPA) (Hepting and Maciag 2005), the other based on a tool developed by one of the authors called cogito (Hepting 2002). Both support tools enabled comparisons of 29 environmentally preferable cleaning products using eight product attributes, described in Table 1 (US-EPA 1997). 56 participants were recruited for the evaluation from the University of Regina Computer Science Participant Pool (Hepting 2006).

Attribute	Definition
Skin irritation (skin)	Amount of redness or swelling of the skin caused by the product. Skin irritation values include: exempt, negligible-slight, slight, medium, strong, not reported. Lower values of skin irritation are preferable
Food chain exposure (fce)	The amount of the product that has potential to seep into the food chain and be consumed by aquatic plants and animals. Values of fce include: exempt, less than 5000, less than 1000, less than 15000, greater than 15000, not reported. Lower fce values are preferred
Air pollution potential (air)	The amount of volatile organic compounds (VOCs) in the product. VOC values include: n/a(0%), less than 1%, less than 5%, less than 15%, less than 30%, greater than 30%, not reported. Lower VOCs are preferred
Product contains fragrance (frag)	Fragrances added to the product. Products with no fragrances are preferred as addition of fragrances increases potential pollution
Product contains dye (dye)	Dyes added to the product. Products with no dye are preferred as addition of dyes increases the risk of pollution
Product is a concen-trate (con)	Product uses reduced packaging. Concentrated packaging is preferred as it limits potential pollution
Product uses recycla-ble Packaging (rec)	Product uses recyclable/recycled paper packaging. This is preferred as it limits potential pollution
Product minimizes exposure to concentrate (exp)	Product is not a concentrate. Although preferred for health reasons, as it limits exposure to the product, it increases potential pollution

Table 1. Eight cleaning product attributes provided by the support tools. Definitions and individual attribute ranges are provided.

The authors are highly interested in evaluating the utility of such tools. In the preliminary evaluation described in (Maciag, 2005) the participants were asked to perform a series of questions on the two support tools described. Participant response times and task scores were noted. User response times were measured to indicate the duration required by each participant in answering prescribed questions using the different tools whereas task scores were measured to indicate the success at which participants were able to answer the prescribed questions.
Based on results obtained, participants were more time and task effective while using the cogito tool developed by the authors. Although these results were encouraging, they explain little with respect to how well consumers would actually perform on these tools for their own purposes. The authors hypothesize that an analysis of user

decision accuracy may have a greater indication of consumer satisfaction and may provide a clearer illustration of the utility of provided tools.

Some preliminary analysis has already been conducted by the authors and described in (Maciag and Hepting 2007) in such regard. The evaluation described in this paper compliments this work and seeks to further examine and evaluate knowledge obtained when conducting this type of analysis.

3 Evaluation Design

As part of the preliminary usability evaluation described in (Maciag 2005) each of the 56 participants were asked to rank the eight product attributes (in Table 1) according to perceived importance using a four point scale: *unimportant*, *somewhat important*, *important*, *very important*. Each participant was also asked to select a product they would consider using for personal applications while using either the US-EPA or *cogito* tool. This information was used as part of the criteria for the procedure described in this paper. To evaluate the procedure, the authors conducted a train and test method. Figure 1 illustrates the details of this procedure.

First, the 29 cleaning products were clustered into four product groups, as described in (Maciag et al. 2007). In order to conduct a proper analysis of product attribute values and specified user rankings, each individual product attribute value needed to be re-coded and mapped to corresponding user rankings. In this preprocessing phase, all data was re-coded into binary, as described in Table 2. For example, if a participant ranked skin irritation highly (important or very important), according to the re-coding procedure in Table 2, products with lower skin irritation (slight or less) would be considered preferred by this participant. Thus, a comparative analysis could be conducted between product attributes and participant rankings to determine the associative strength between them.

Using the product sample as the training set, a decision system was constructed consisting of the 29 cleaning products, their re-coded attributes values (Table 2), and their cluster membership values (1-4) as per the previous results obtained in (Maciag et al. 2007). Rough set reduction techniques were then performed on the training sample to reduce those attributes with similar values across the obtained product clusters. Next, the 56 participants were split into two separate testing sets based on the support tool used when selecting a cleaning product. 28 participants were delegated to each testing set accordingly. Since some participants did not select a product in the evaluation (10 and 4 participants did not select a product when using the US-EPA and *cogito* tool respectively), these participants were omitted from further evaluation. Both testing sets were evaluated using the reduct results obtained from the training set. Results describing classification accuracy and coverage were analyzed to determine a measurement of utility for each tool.

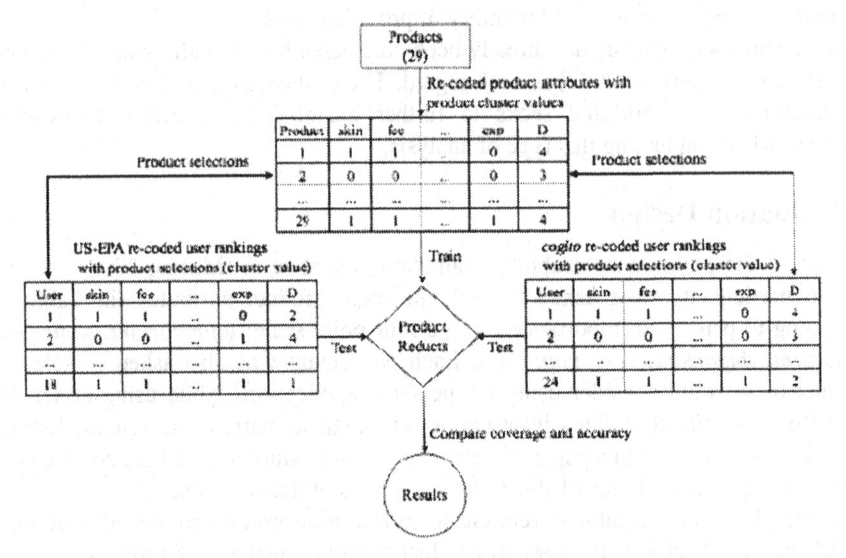

Figure 1. Illustration of the evaluation procedure. Initially product attributes and user rankings are re-coded to indicate their associative strength. The product set is used as a training set where reducts are formulated. Testing samples, comprised of re-coded participant rankings and product cluster values based on participant selections, are constructed based on the specific tools used and are each tested with the reducts generated by the product set accordingly.

Attribute/User Ranking	Important/ Very Important
skin	slight or less
fce	less than 5000
air	less than 1%
frag	no
dye	no
con	yes
rec	yes
exp	no

Table 2. Description of the re-coding procedure used in the analysis. Product attributes and user rankings were recoded into binary and mapped according to the described ranges to indicate their associative strength. In the analysis, rankings of important and very important were re-coded as "1" and "0" otherwise. Similarly, those attributes within the described ranges corresponding to these rankings were re-coded as "1" and "0" otherwise.

4 Results and Discussion

Based on the rough set reduction procedure in the second phase of the evaluation (on the product sample), two reducts each with five attributes were generated: {skin, fce, air, dye, con} and {skin, fce, air, dye, exp}, indicating three of the eight attributes across product clusters contained similar attribute values. Each reduct generated had a positive region of 86.2%. Since the positive region was not exactly 100%, this may indicate that there exist some cleaning products among the four clusters generated that were difficult to cluster. This may have been the result of the attribute recoding process described in Table 2 or it could be that although certain products were assigned to different clusters, it may be that the distance between a select few of these products is minimal, e.g. products in opposing clusters could be relatively close in proximity. This analysis is left for future evaluation.

US-EPA Train/Test Results							
Actual	**Predicted**				**#Objects**	**Accuracy**	**Coverage**
	3	4	1	2			
3	0	0	0	0	1	0	0
4	0	13	0	2	17	86.7%	88.2%
1	0	0	0	0	0	0	0
2	0	0	0	0	0	0	0
Totals					**18**	**86.7%**	**83.3%**
cogito Train/Test Results							
	3	4	1	2			
3	0	0	0	0	1	0	0
4	1	16	0	2	22	94.1%	77.3%
1	0	0	0	0	0	0	0
2	0	0	0	0	1	0	0
Totals					**24**	**94.1%**	**70.8%**

Table 3. Results of the evaluation. Results indicated refer to those participants who were correctly classified according to assigned product clusters based on the train and test procedure. Accuracy and coverage measures are used to indicate the percentage of correctly classified users as well as to indicate a measurement of decision accuracy and utility.

Table 3 provides the results of the training and testing procedure. The authors hypothesize that decision accuracy could be analyzed from the perspective of obtained accuracy and coverage. Based on results indicated in Table 3, when observing the results for accuracy totals, it would appear that participants obtained slightly higher decision accuracy when using the *cogito* tool. However, when observing the result for coverage, results indicate that some participants could not be classified properly, or at all, which may indicate that these participants performed ineffectively. There is a similar illustration when observing the results obtained for the US-EPA tool. Although classification accuracy is slightly less than that obtained from participants using the *cogito* tool, coverage is higher indicating that slightly more participants were classified properly.

In either case, based on the knowledge obtained from this analysis, even though it may seem that participants who used the *cogito* tool did obtain slightly higher decision accuracy based on accuracy totals, it would seem that each support tool would benefit from improvements made to their interfaces since neither tool obtained complete accuracy or coverage percentages. Future work will include a more detailed analysis of the relationship between accuracy and coverage measurements and how they correspond with measurements of user decision accuracy.

5 Conclusion

The primary objective of this paper was to describe a procedure to evaluate the utility of web-based consumer support tools by measuring decision accuracy using techniques in rough sets. Based on results obtained, techniques in rough sets may be used to obtain knowledge of web usage patterns as results from the evaluation yielded interesting indicators on the overall utility of the tools used in the evaluation.

Although the tools used were in the domain of environmentally preferable purchasing, it is hypothesized that online retailers, regardless of what retail domain they conduct business, could utilize the described procedure to conduct the type of analysis described in this paper.

Results indicated that both tools evaluated could improve utility scores by incorporating enhancements in their design. One aspect of design enhancement that may be considered includes personalization. In this respect, the authors have already conducted research, described in Maciag et al. (2007), into personalizing aspects of the user interfaces in such tools. The authors hypothesize that personalization may greatly increase user decision accuracy, thus increase the utility of these tools.

The authors are also currently researching new techniques to incorporate automatic decision accuracy feedback in such tools, thus enabling consumers to immediately visualize how they could improve upon current selections. This type of functionality may increase the utility of these tools, as consumers would be able to visualize the quality of their selections immediately upon selection and revise accordingly. Future work will include implementation and evaluation of described procedures.

References

Beaubouef, T., Ladner, R., and Petry, F. (2004) *Rough Set Spatial Data Modeling for Data Mining.* International Journal of Intelligent Systems, Vol. 19, pp. 567-584.

Chimphlee, S., Salim, N., Ngadiman, M.S.B., Chimphlee, W., Srinoy, S. (2006) *Independent Component Analysis and Rough Fuzzy Based Approach to Web Usage Mining.* In Proc. International Multi-Conference on Artificial Intelligence and Applications (IASTED), pp. 422-427.

Eirinaki, M. and Vazirgiannis, M. (2003) *Web Mining for Web Personalization.* ACM Transactions on Internet Technology, Vol.3, No.1, pp.1-27.

Goebel, M. and Gruenwald, L. (1999) *A Survey of Data Mining and Knowledge Discovery Software Tools.* SIGKDD Explorations, Vol.1, pp. 20-33.

Ha, S.H. (2002) *Helping Online Customers Decide Through Web Personalization.* In Proc. IEEE Intelligent Systems, pp.34-43.

Hepting, D.H. (2002) *Towards a Visual Interface for Information Visualization.* In Proc. IEEE International Conference on Information Visualization, pp. 295-302.

Hepting, D.H. and Maciag, T. (2005) *Consumer Decision Support for Product Selections.* In Proc. International Symposium on Environmental Software Systems (ISESS).

Hepting, D.H. (2006) *Ethics and Usability Testing in Computer Science Education.* ACM SIGCSE Bulletin, Vol.38, No.2, pp. 76-80.

Holland, S., Ester, M., and KieBling, W. (2003) *Preference Mining: A Novel Approach on Mining User Preferences for Personalized Applications.* In Proc. European Knowledge Discovery and Databases (PKDD), pp.204-216.

Kosala, R. and Blockeel, H. (2000) *Web Mining Research: A Survey.* SIGKDD Explorations, Vol.2, pp.1-15.

Lingras, P. and West, C. (2004) *Interval Set Clustering of Web Users with Rough K-Means.* Journal of Intelligent Information Systems, Vol.23, pp. 5-16.

Maciag, T. (2005) *An Evaluation of User Interface for Environmental Decision Support Systems.* University of Regina (Masters thesis).

Maciag, T., Hepting, D.H., Slezak, D., and Hilderman, R.J. (2007) *Mining Associations for Interface Design.* To appear in Proc. Joint Rough Set Conference.

Maciag, T. and Hepting, D.H. (2007) *Discovery of Usability Patterns in Support of Green Purchasing.* To appear in Proc. International Symposium on Environmental Software Systems (ISESS).

Pawlak, Z. (1991) *Rough Sets, Theoretical Aspects of Reasoning About Data.* Kluwer Academic Publishers.

Pierrakos, D., Paliouras, G., Papatheodorou, C., and Spyropoulos, C.D. (2003) *Web Usage Mining as a Tool for Personalization: A Survey.* User Modeling and User-Adapted Interaction, Vol.13, pp.311-372.

Pu, P. and Chen, L. (2005) *Integrating Tradeoff Support in Product Search Tools for Ecommerce Sites.* In Proc. Electronic Commerce (EC), pp.269-278.

Slezak, D. (2005) *Association Reducts: A Framework for Mining Multi-Attribute Dependencies.* In Proc. International Symposium on Methodologies for Intelligent Systems (ISMIS), pp. 354-363.

Spiliopoulou, M. (2000) *Web Usage Mining for Web Site Evaluation.* Communications of the ACM. Vol.43, No.8, pp.127-134.

Srivastava, J., Cooley, R., Deshpande, M., and Tan, P.N. (2000) *Web Usage Mining: Discovery and Applications of Usage Patterns from Web Data.* SIGKDD xplorations, pp.12-23.

United States Environmental Protection Agency (1997) *Cleaning Products Pilot Project Fact Sheet.* http://www.epa.gov/opptintr/epp/pubs/cleanfct.pdf (Accessed March 2007).

Zhou, B., Hui, S.C., Chang, K. (2004) *A Formal Concept Analysis Approach for Web Usage Mining.* In Proc. Intelligent Information Processing 2, pp.437-441.

Local Table Condensation in Rough Set Approach for Jumping Emerging Pattern Induction*

Pawel Terlecki and Krzysztof Walczak

Institute of Computer Science, Warsaw University of Technology, Nowowiejska 15/19, 00-665 Warsaw, Poland
P.Terlecki, K.Walczak@ii.pw.edu.pl

Abstract. This paper extends the rough set approach for JEP induction based on the notion of a condensed decision table. The original transaction database is transformed to a relational form and patterns are induced by means of local reducts. The transformation employs an item aggregation obtained by coloring a graph that re°ects con°icts among items. For e±ciency reasons we propose to perform this preprocessing locally, i.e. at the transaction level, to achieve a higher dimensionality gain. Special maintenance strategy is also used to avoid graph rebuilds. Both global and local approach have been tested and discussed for dense and synthetically generated sparse datasets.

Key words: jumping emerging pattern, transaction database, local reduct, locally condensed decision table, rough set, graph coloring

1 Introduction

The majority of knowledge discovery problems is general and remains similarly deffined for various types of data. Classiffication, association ffinding, clustering, pattern induction are only several examples of tasks considered for relational, transactional or temporary databases. On the other hand, due to data speciffiicity discovery algorithms are often signifficantly di®erent. Here, we look at the problem of JEP induction that is originally a data mining issue. A JEP is a pattern in a classiffied dataset, supported in one class and absent in other classes. This kind of patterns has proved to have a highly discriminative power ([1]) and have been successfully applied to classiffication of business and biological data ([2]).
Our paper follows the idea of building a condensed decision table for a given transaction database. While in our previous work ([5]), we have proposed a global approach, which sought an optimal proper aggregation of items taking into account all transactions, here, a local method has been put forward. For each transaction, we focus on these transactions that are important in terms of building an indiscernibility relation for the respective object. By considering fewer transactions, we end up with better transaction-wise aggregations and, thus, higher problem dimensionality gains.

* The research has been partially supported by grant No 3 T11C 002 29 received from Polish Ministry of Education and Science.

Each aggregation (partition) can be found by solving a graph coloring problem in a graph expressing which item pairs can be in the same block. In order to avoid a signifficant overhead caused by constructing graphs from scratch, we presort transactions, such that we can keep one graph and perform relatively small modiffications to its edges while the algorithm proceeds. Both, global and local approaches have been discussed and compared over dense and syntactically generated sparse transactional data.
Section 2 covers the theoretical background on emerging patterns and the rough set theory. In Sect. 3, we derive the idea of local condensation from the global approach. The novel algorithm for JEP induction is given in Sect. 4. Section 5 presents an experiment. The paper is summarized in Sect. 6.

2 Preliminaries

Let a transaction system be a pair $(\mathcal{D}; \mathcal{I})$, where $\mathcal{D}$ is a finite sequence of transactions $(T_1;...; T_n)$ (database) such as $T_i \subseteq \mathcal{I}$ for $i = 1; :::; n$ and $\mathcal{I}$ is a non- empty set of items (itemspace). A support of an itemset $X \subseteq \mathcal{I}$ in a sequence $D = (T_i)_{i \in k} \subseteq D$ is deffined as $supp_D(X) = \frac{|\{i \in k : x \subseteq T_i\}|}{|k|}$, where $K \subseteq \{1; ...; n\}$.

Let a decision transaction system be a tuple $(\mathcal{D}; \mathcal{I}; \mathcal{I}_d)$, where $(\mathcal{D}; \mathcal{I} \cup \mathcal{I}_d)$ is a transaction system and $\forall_{T \in D} |T \cap I_d| = 1$. Elements of $\mathcal{I}$ and $\mathcal{I}_d$ are called condition and decision items, respectively. A support for a decision transaction system $(\mathcal{D}; \mathcal{I}; \mathcal{I}_d)$ is understood as a support in the transaction system $(\mathcal{D}; \mathcal{I} \cup \mathcal{I}_d)$.

For a decision item $c \in \mathcal{I}_d$, we deffine a decision class sequence $C_c = (T_i)_{i \in k}$, where $K = \{k \in \{1; ...; n\} : c \in T_k\}$. Notice that each of the transactions from $\mathcal{D}$ belongs to exactly one class sequence. For a database $\mathcal{D} = (T_i)_{i \in k \subseteq \{1,...,n\}} \subseteq \mathcal{D}$, we deffine a complement database $\mathcal{D}' = (T_i)_{i \in \{1,...,n\}-k}$. Given two databases $\mathcal{D}_1; \mathcal{D}_2 \subseteq \mathcal{D}$ we deffine a jumping emerging pattern (JEP) from $\mathcal{D}_1$ to $\mathcal{D}_2$ as an itemset $X \subseteq \mathcal{I}$ such as $supp_{D1}(X) = 0$ and $supp_{D2}(X) > 0$. A set of all JEPs from $\mathcal{D}_1$ to $\mathcal{D}_2$ is called a JEP space and denoted by $JEP(\mathcal{D}_1, \mathcal{D}_2)$.
JEP space can be represented in a concise manner [1]. Consider a set S. A border is an ordered pair $< \mathcal{L}, \mathcal{R} >$ such that $\mathcal{L}, \mathcal{R} \subseteq 2^S$ are antichains and $\forall_{X \in L} \exists_{Z \in R} X \subseteq Z$. $\mathcal{L}$ and $\mathcal{R}$ are called a left and a right bound, respectively. A border $< \mathcal{L}, \mathcal{R} >$ represents a set interval $[\mathcal{L}, \mathcal{R}] = \{Y \in 2^S : \exists_{X \in L} \exists_{Z \in R} X \subseteq Y \subseteq Z\}$. Now, consider a decision transaction database $(\mathcal{D}; \mathcal{I}; \mathcal{I}_d)$ and two data- bases $\mathcal{D}_1; \mathcal{D}_2 \subseteq \mathcal{D}$. According to [1], $JEP(\mathcal{D}_1; \mathcal{D}_2)$ can be uniquely represented by a border. For $d \in I_d$, we use a border $< \mathcal{L}_d; \mathcal{R}_d >$ to describe the JEP space $JEP(C'_d; C_d)$. Members of left bounds are minimal JEPs.
Basic notions from the rough set theory follow the deffinitions from [3].

3 Transaction Database Condensation

Global Condensation. In [5], we put forward an idea of a condensed table for a decision transaction system. Without loss of information on JEPs, items are aggregated and transformed into attributes of a decision table, so that the ffinal structure is a concise relational description of classiffied transactional data.
Hereinafter, we assume that our data is given by a decision transaction system $DTS = (\mathcal{D}, \mathcal{I}, \mathcal{I}_d)$, where $D = (T_1, ..., T_n)$, $\mathcal{I} = \{I_1, ..., I_m\}$, $\mathcal{I}_d = \{c_1, .., c_p I\}$.

We say that a partition $\{p_1, ..., p_r\}$ of $\mathcal{I}$ is proper if $\forall_{T \in D} \forall_{j \in \{1,...,r\}} \left| T \cup P_j \right| \leq 1$. A (globally) condensed decision table for: *DTS*, a decision transaction system, $P = \{p_1, ..., p_r\}$, a proper partition of $\mathcal{I}$, $F = \{f_1, ..., f_r\}$, where $f_j : 2^{P} i \mapsto \mathbf{N}$ and f_j is a bijection for each $j \in \{1, ..., r\}$ is a decision table $CDT_{DTS,P,F} = (\mathcal{U}, \mathcal{C}, d)$ such that $\mathcal{U} = \{u_1, \ldots, u_n\}$, $\mathcal{C} = \{a_1, ..., a_r\}$, $V_d = \{d_1, ..., d_p\}$; $a_j(u_i) = f_j(T_i \cup p_j)$, $\forall_{i \in 1,...,n, j \in 1,...,r}; d(u_i) = T_i \cap I_d, \forall_{i \in 1,...,n}$. For convenience, we introduce the notation: $condPatt_{DTS;P;F}$ $(u; B) = \bigcup_{k \in K} f_k^{-1}(ak(u))$, where $u \in \mathcal{U}$, $B = \{ak\}_{k \in K \subseteq \{1,...,r\}}$. When it does not lead to unambiguity, indices for *CDT* and *condPatt* will be omitted for brevity. Since a mapping function vector *F* do not change the structure of a decision table in terms of row discernibility, we use one type of mapping, referred as natural, in which each function maps a new itemset into a next natural number, while moving in the order of transaction indices.
It has been proved in [5] that information on JEPs is preserved after condensation. A minimal JEP can be obtained from at least one object from a positive region of a condensed table after combining with one of its local reducts.

Table 1. A sample decision transaction system DTS = $\{\{T_1, ..., T_6\}, \{c_0, c_1\}\}$, a respective binary table and a condensed table for the partition $\{\{c, d, e\}, \{f, h\}, \{a, i\}, \{b, g\}\}$ and the natural *F*.

T_1	di	c_0
T_2	$befi$	c_0
T_3	aeg	c_0
T_4	ch	c_1
T_5	eh	c_1
T_6	bi	c_1

$\Longrightarrow$

	a	b	c	d	e	f	g	h	i	d
u_1	0	0	0	1	0	0	0	0	1	0
u_2	0	1	0	0	1	1	0	0	1	0
u_3	1	0	0	0	1	0	1	0	0	0
u_4	0	0	1	0	0	0	0	1	0	1
u_5	0	0	0	0	1	0	0	1	0	1
u_6	0	1	0	0	0	0	0	0	1	1

$\Longrightarrow$

	a_1	a_2	a_3	a_4	d
u_1	0	0	0	0	0
u_2	1	1	0	1	0
u_3	1	0	1	2	0
u_4	2	2	2	0	1
u_5	1	2	2	0	1
u_6	3	0	0	1	1

Local Condensation. In the text, we focus on the problem of e±cient pattern computation rather than on providing an alternative data representation. According to the deffinition a pattern is a JEP if it is supported in one class and is not supported in others. In our approach each transaction is considered individually in order to compute all minimal JEPs that it supports. Notice that this step requires information on the considered transaction and on all transactions from other classes. The rest of transactions from the same class do not play any role. Thus, instead of analyzing one condensed table for a decision transaction system, one can consider a separate condensed table for each transaction.

A locally condensed table for: *DTS*, a decision transaction system, $P = \{p_1,..., p_r\}$, a proper partition of $\mathcal{I}$, $F = \{f_1,..., f_r\}$, where $f_j : 2^{p_j} \mapsto \mathbf{N}$ and f_j is a bijection for each $j \in \{1,...,r\}$ and $T_i \in D$, where $i = 1,..., |D|$, a transaction is a condensed decision table $LCDT_{DTS;P;F;T_i} = CDT_{DTS_i;P;F}$, where $DTS_i = (D^i; I; I_d)$, $D^i = \{T_i; T_k\}_{k \in K}$ and $K = \{k \in \{1,..., |D|\} : k \neq i \wedge T_k \cap I_d \neq T_i \cap I_d\}$.

We also deffine $locCondPatt_{DTS;P;F;T_i}(B) = condPatt_{DTS_i;P;F}(u_1; B)$.

Table 2. Locally condensed decision tables $LCDT_{DTS;P_i;F_i;T_i}$ for $i = 1; 2; 3$, $P_1 = \{\{a; f; g; h; i\}; \{b; c; d; e\}\}$, $P_2 = \{\{a; c; e\}; \{b; h\}; \{d; f\}; \{g; i\}\}$; $P_3 = \{\{b; c; d; e\}; \{a; h; i\}; \{f; g\}\}$ and $F_1; F_2; F_3$ deffined in a natural way.

P_1, F_1, T_1			
	a_1^1	a_2^1	d
u_1	0	0	0
u_4	1	1	1
u_5	1	2	1
u_6	0	3	1

P_2, F_2, T_2					
	a_1^2	a_2^2	a_3^2	a_4^2	d
u_2	0	0	0	0	0
u_4	1	1	1	1	1
u_5	0	1	1	1	1
u_6	2	0	1	0	1

P_3, F_3, T_3				
	a_1^3	a_2^3	a_3^3	d
u_3	0	0	0	0
u_4	1	1	1	1
u_5	0	1	1	1
u_6	2	2	1	1

4 JEP Induction

Algorithm Overview. Let us assume that data is given by a decision transac- tion system be a tuple DTS = $(\mathcal{D}; \mathcal{I}; \mathcal{I}_d)$, where $\mathcal{D} = (T_k)_{k \in \{1,..., |D|\}}$. The goal is to induce minimal patterns $\mathcal{L}_c$ from the class C_c' to C_c for each c $\in I_d$.

For each transaction, a good item aggregation is computed, locally condensed table is constructed and a set of local reducts for a respective object is induced. Reducts are used to generate minimal patterns, which are added to the left bound of the transaction's class. Each aggregation is a solution to a graph coloring problem for an item-conflict graph $G_{DTSk} = (V; E_k)$, where $V = \{v_1,.., v_m\}$,

$\forall x, g \in \{1,...,m\}(v_x, v_g) \in E \Leftrightarrow \forall T \in D^{ki_x} \notin T \vee i_y \notin T$ for each k=1...n ([5]).

1: $\mathcal{L}c = \phi$; for each c $\in I_d$, assume GDTS;T_0 = (V; ϕ), $T_0 \cap I_d$

2: **for** (k = 1; 1 <= $|D|$; k + +) **do**

3: Prepare a local item-conflict graph G_{DTSk} using G_{DTSk-1}

4: Find a good coloring P_k in $GDTS_k$

5: Construct a locally condensed decision table $LCDT_{DTS;Pk;Fk;Tk}$, F_k - natural

6: Compute REDLOC(u_1; d)

7: $\mathcal{L}_c = \mathcal{L}_c \cup \{locCondPatt_{DTS;Pk;Fk;Tk}(u_1; R) : R \in REDLOC(u_1; d)\}$, $c = T_k \cup I_d$
8: **end for**

Good Partition Finding. Condensation is a way to reduce the problem's dimensionality and does not affect the completeness of the ffinal result. Therefore, we are satisffied with suboptimal solutions obtained with heuristics, like LF, SLR, RLF, SR ([5]). Unlike for a global approach, here, a locally condensed decision table is constructed for each transaction. For large databases graph construction good coloring ffinding and condensed table rebuild can add a signifficant overhead to computation. To diminish the impact of the ffirst factor, we avoid complete graph reconstructions. For maintenance reasons, let us deffine, for each G_{DTSk}, a function $h_k : E_k \mapsto \mathbf{N}$, where $f(e_{xy}) = |\{T \subset D^k : x < y \wedge (i_x \notin T \vee i_g \notin T)\}| > 0$ for $e_{xy} = (u_x; u_y) \in E_k$, i.e. h_k re°ects the number of transactions for which items cannot be aggregated. For k=1, we have to build a complete graph. Assuming that we have G_{DTS} and h_{k-1}, we have two scenarios. First, when both T_{k-1}, T_k belong to the same class, we do not consider transactions from other classes, we have $E_k = E_{k-1} - \{e_{xy} \in E_{k-1} : h_{k-1}(e_{xy}) = 1 \wedge i_x, i_y \in T_{k-1}\} \cup \{e_{xy} \notin E_{k-1} : i_x, i_y \in T_k\}$ and h_k is updated similarly. In the second case, T_{k-1}, T_k belong to the di®erent classes d_{k-1}, d_k, respectively. Here, we want to affect transactions from other classes, i.e.

$$E_k = \{e_{xy} \in E_{k-1} : |\{T \in C_{d_k - T_k} : i_x, i_y \in T\}| > h_{k-1}(e_{xy})\} \cup \{e_{xx} \notin E_{k-1} : \exists_{T \in C_{d_{k-1}}} i_x, i_y \in T\}$$

Since the ffirst situation is typically more beneffícial, we sort transactions in $\mathcal{D}$ by class.

5 Experimental Results

The novel local method has been compared with our, previously proposed, global approach. JEP induction methods are usually tested over dense datasets originating from UCI Repository ([6]). Since sparse data remain a typical interest of KDD, we have decided to extend our experiment to synthetically prepared sparse data. Transaction databases have been obtained by means of IBM generator ([7]). Then, the data have been clustered with a CLUTO package ([8]). We favored less numerous clusterings with possible similar sizes of classes. Reducts have been computed by means of the exact method ([3]), complete view approaches with candidate pruning were ine±cient for considered dimensionality.

As we can see in Tab. 3, for sparse datasets the local approach solves problems of lower dimensionality. The time di®erence in preprocessing between both methods is noticeable; however, total time is typically smaller by one order of magnitude. On the other hand, for dense data it is very unlikely to get better local aggregations and the global algorithm is faster by a preprocessing overhead.

6 Conclusions

In this paper a novel rough set approach to JEP induction has been put forward. Following the idea of a globally condensed decision table, that is a concise representation of a given classiffied transactional dataset obtained by item aggregation, we have introduced local condensation performed for each individual transaction. In this approach, only transactions from classes different that a given transaction is considered, which gives signifficantly better aggregations? Since aggregation ffinding involves solving a graph coloring problem, the higher dimensionality gain is traded off for constructing and analyzing graphs per each transaction.

Table 3. Experimental results. For global and local approach, dimensionality, pre-processing (steps: 3, 4, 5) and pattern computation time (6, 7). D and I given in 1000s.

Dataset	JEPs	Global Attr	Global PrepTime	Global Patt Time	Local Attr	Local PrepTime	Local Patt Time
balance	303	4.00	63	421	4.00	2838	424
car	246	6.00	62	2594	6.00	21238	2542
cmc	2922	9.00	78	10188	9.00	28927	9858
geo	7458	10.00	94	2265	10.35	9242	1512
house	6986	16.00	78	22187	16.00	3814	18844
lymn	6794	18.00	78	10750	18.00	1002	10512
vehicle	22070	20.00	110	97625	19.74	37735	84515
D0.3 I0.2 T7 C3	3370	27.00	187	1023890	21.97	24762	159300
D0.4 I0.1 T5 C2	1601	25.00	109	350000	18.07	9034	23793
D1.0 I0.07 T3 C2	1361	25.00	125	466750	17.58	41027	58535
D1.1 I0.3 T4 C2	4434	26.00	1110	1552625	20.42	633375	122785
D2.5 I0.03 T3 C3	1654	28.00	156	701110	24.18	110018	577202
D4.0 I0.015 T2 C2	186	14.00	141	89094	13.00	145161	75691
D6.0 I0.012 T1 C2	12	1.00	125	5656	1.00	94873	3336

To decrease this overhead, we order transactions by class and extend information stored with each graph, so that it does not have to be rebuilt from scratch but obtained from the previous one. Global and local condensations have been compared for real and synthetic datasets. Experiments have shown that for sparse data, which is a typical case in KDD, local approach highly reduces dimensionality and computation time. However, global approach is better for dense datasets, when local aggregations are comparable to the global solution.

References

G. Dong and J. Li, \Mining border descriptions of emerging patterns from dataset pairs," *Knowl. Inf. Syst.*, vol. 8, no. 2, pp. 178{202, 2005.

J. Li and L. Wong, \Emerging patterns and gene expression data," in *Proc. of 12th Workshop on Genome Informatics*, pp. 3{13, 2001.

J. Bazan, H. S. Nguyen, S. H. Nguyen, P. Synak, and J. Wroblewski, \Rough set algorithms in classiffication problem," *Rough set methods and applications: new de- velopments in knowl. disc. in inf. syst.*, pp. 49{88, 2000.

P. Terlecki and K. Walczak, \Local reducts and jumping emerging patterns in relational databases," vol. 4259 of *LNCS*, pp. 268{276.

P. Terlecki and K. Walczak, \Jumping emerging pattern induction by means of graph coloring and local reducts in transaction databases," in *JRS'07*, 2007.

C. B. D.J. Newman, S. Hettich and C. Merz, \UCI repository of machine learning databases," 1998.

R. Agrawal and R. Srikant, \Fast algorithms for mining association rules in large databases," in *VLDB '94*, pp. 487{499, 1994.

G. Karypsis, *CLUTO. A Clustering Toolkit. Release 2.0.* Univ. of Minnesota, 2002.

Rough Monadic Interpretations of Pharmacologic Information

P. Eklund[1], *M.A. Gal'an*[2*], *and J. Karlsson*[3]

1 Department of Computing Science, Umea University, Sweden,
email: peklund@cs.umu.se
2 Department of Applied Mathematics, University of M'alaga, Spain,
email: magalan@ctima.uma.es
3 Department of Computer Architecture, University of M'alaga, Spain,
email: johan@ac.uma.es

Abstract. Public databases for pharmacological information provide rich and complete information for therapeutic requirements. In particular, the ATC code with its unique identification of drug compound is the basis e.g. of modelling of drug interactions. Therapeutic decisions are frequently faced with needs to consider not just drug-drug interactions but also drugs interacting with sets of drugs and even sets of medical conditions. The typing of interactions provides an additional complication as this is an entrance towards management of uncertainties concerning interactions. The hierarchical structure of pharmacological information and its intrinsic uncertainties are in this paper used as an information platform for the demonstration of the power of rough set modelling. Rough sets in a more general functorial view involving partially ordered monads are able to capture interactions with respect to different granularities in the information hierarchy.

1 Introduction

The data structure for pharmacologic information is hierarchical in its tree-like subdivision from anatomic, through therapeutic down to chemical information of the drug compound. The data is also clustered given the chemical nature of the drug and its relations to medical conditions. Information systems and national catalogues of drugs aim at being complete with respect to chemical declarations, indications / contraindications, warnings, interactions, side-effects, pharmacodynamics / pharmacokinetics, and pure pharmaceutical information. This in connection with the ATC code and package identification codes provides a well-defined information structure and complete information source for therapeutic and various statistical needs.

We will in this paper use rough sets, and generally rough monads, for decision modelling in particular for problems related to interactive drugs. Apart from drug-drug interactions we will also discuss possibilities of modelling drug-disease interactions. In these approaches we need to move beyond traditional rough sets.

*Partially supported by Spanish projects TIC2003-09001-C02-01 and TIN2006-15455-C03-01.

Indeed, we will provide structural generalization of what Pawlak [14] calls information systems, i.e. tables involving objects of a universe together with a set of attributes identifying each object. Our categorical framework was presented in [1], where we use partially ordered monads to generalize the ordinary powerset functor view of relations to enabling the use of a wide range of set functors for representing various hierarchies involving sets.
The paper is organized as follows. Section 2 gives an introduction to the structure of pharmacological information. In Section 3, rough monads are introduced with respect to interpretations of pharmacological structures given in Section 4. Section 5 concludes the paper.

2 Pharmacological information

The Anatomic Therapeutic Chemical (ATC) classification system is an accepted WHO standard. For drug utilization statistics, a unit of measurement called defined daily dose (DDD) has also been developed to complement ATC. A DDD is the average dose per day for a drug that is used for its main indication when treating adults. This is not to be confused with the guideline or recommendation of dosage. Indeed, the DDD could be in the middle between two commonly prescribed dosages and as such never be an actual prescribed dosage.
To outline the ideas in ATC, we look in Table 1 on the classification of verapamil (code C08DA01) for hypertension with stable angina pectoris.

C	cardiac and vessel disease medication	1st level main anatomical group
C08	calcium channel blockers	2nd level the rapeutic subgroup
C08D	selective cardiac calcium channel blockers	3rd level pharmacological subgroup
C08DA	phenylalcylamins	4th level chemical subgroup
C08DA01	verapamil	5th level

Table 1. Classification of *verapamil.*

The DDD for verapamil is 240-480 mg/day divided on three dosage occasions. The administration routes registered are oral and injected.
It is important to note that the drugs in ATC are, with a very few exceptions, classified according to their main indication of use. As such, the ATC terminology system should not be used as a recommendation of therapeutic usage. Drugs in the same ATC group are not necessarily therapeutically equivalent (generic substitutes), even if they are classified under the same chemical subgroup (4th level ATC). Drugs with the same ATC code can have different dosages and administration routes.
Note also that the ATC coded is for therapeutic use, while the article code is a unique identifier which is useful in patient journal notes, inventory systems, and so on.
The quantification of degree of interaction is unclear. Qualifications have, however, been suggested. For drugs showing therapeutically significant interactions, it is important to distinguish between different types of interactions, like recommended combination, neutral combination (no harmful interactions), risky combination (should be monitored), dangerous combination (should be avoided) and possible

interaction (not tested). In this qualification it is clear that a corresponding linear quantification is not straightforward.
Further, the drugs are affected in different ways, according to *no change in effect*, *increases effect*, *reduce effect* and *other* (e.g. a new type of side effect). Interaction type and effect need to be considered in the guideline for respective treatments.

3 Rough monads

Rough sets and fuzzy sets are both methods to represent uncertainty. By using partially ordered monads we can find connections between these two concepts. Considering the partially ordered powerset monad, we showed in [1] how rough sets operations can be provided in order to complement the many-valued situation. This is accomplished by defining "rough monads".
Partially ordered monads, see [4, 1] for definitions and examples, are appropriate categorical formalizations and generalizations of rough sets. A relation R on X corresponds to a mapping $\rho x: X \longrightarrow PX$, where $\rho x(x) = \{y \in X | xRy\}$ and the inverse relation R^{-1} is represented as $\rho_x^{-1}(x) = \{g \in X | xR^{-1}y\}$. This is the partially ordered monadic reformulation of rough sets based on the powerset partially ordered monad [1].
For a partially ordered monad $(\varphi, \leq, \eta, \mu)$, let $\rho x: X \longrightarrow \varphi X$ be a corresponding generalized relation, and let $a \in \varphi X$. The inverse must be specified for the given set functor φ. The upper and lower approximations are then

$$\uparrow X(a) = \mu X \circ \varphi \rho_x^{-1}(a) \quad \downarrow X(a) = \bigvee_{\rho X(x) \leq a} \eta X(x)$$

with the monadic generalizations of ρ-weakenedness and ρ-substantiatedness, for $a \in \varphi X$, being

$$\Uparrow X(a) = \mu X \circ \varphi \rho X(a) \quad \Downarrow X(a) = \bigvee_{\rho_X^{-1}(x) \leq a} \eta X(x)$$

In the case of $\varphi = P$, i.e. the conventional powerset partially ordered monad; these operators coincide with those for classical rough sets. In this case inverse relations exist accordingly. In the case of of fuzzy sets we use the many-valued powerset partially ordered monad based on the many-valued extension of P to L_{id} as follows. With L a completely distributive lattice, the functor L_{id} is obtained by $L_{id}X$ being the set of mappings $A: X \rightarrow L$. The partial order $\leq$ on $L_{id}X$ is given pointwise. Morphisms $f: X \rightarrow Y$ in Set are extended according to

$$L_{idf} = \bigvee_{f(x)=y} A(x)$$

Finally, $\eta X : X \rightarrow L_{id} X$ and $\mu X : L_{id} X \circ L_{id} X \rightarrow L_{id} X$ are given by

$$\eta X(x)(x') = \{^{1 if X' \leq x}_{0 otherwise} \quad \mu X(M)(x) = \bigvee_{A \in L_{idX}} A(x) \wedge X(A)$$

In case of $\varphi = L^{id}$ the inverse can be given by

$$\rho_X^{-1}(x)(x') = \vee_{y \in X} \rho(x')(x) \wedge \eta X(g)(x')$$

and then, as easily seen $\rho_X^{-1}(x)(x') = \rho(x')(x)$

For other algebraic approaches to rough sets, see e.g. [7, 8]

4 Rough monadic interpretations of drug interactions

In order to study interactions it is sufficient to identify drugs with the same ATC code. In the following, when we speak of drugs we thus refer to an ATC code. We will consider only drug-drug interactions, where medical conditions will specify relational structures of interest.

We draw our case study from guideline based pharmacologic treatment of hypertension [18]. See also [15] for an implementation of these guidelines for primary care.

Typical drugs for hypertension treatment are beta-blockers (C07, C07A) like an atenolol (C07AB03) and diuretics (C03) like thiazides (C03A, C03AA). Atenolol is a selective beta-1-blocker (C07AB). A frequently used thiazide is hydrochlorothiazide (C03AA03). Note that beta-blockers are both therapeutic as well as pharmacological subgroups. Similarly, thiazides are both pharmacological as chemical subgroups.

Providing broad detail of interactions and considerations of medical conditions with respect to pharmacological treatment of hypertension is outside the scope of this paper. However, we can provide a basic example concerning interactions with treatment of diabetes. Beta-blockers may mask and prolong beta-blockers insulin-induced hypoglycemia. If the patients shows the medical condition of diabetes without any other medical conditions present, then the ACE inhibitor (C09A, C09AA) enalapril (C09AA02) is the first choice for treatment. For more detail on interactions and considerations of medical conditions, see [18].

In order to proceed with rough monadic interpretations of drug interactions, let X be the set of drugs, i.e. a set of ATC codes. As a first step, let us consider two-valued interactions, i.e. drug either interact or they don't (recommended or neutral combination). The relation $\rho_X^P : X \to PX$ is given by $\rho_X^P(x)$ being the set of drugs not interacting with drug x. This relation is symmetric in this simplified view of interactions.

Let now A $\subseteq$ X be the set of drugs used for hypertension treatment. Then $A^{\uparrow} = \bigcup_{x \in A} \rho_X^P(x), i.e. A^{\uparrow} (= A^{\Uparrow})$ is the set of all drugs in X that do not interact with any drug used for hypertension treatment. Further, note that there is no hypertension treatment drug which interacts with every drug not used for hypertension treatment. Thus $A^{\downarrow} (= A^{\Downarrow}) = \bigcup_{\rho_X^P(x) \subseteq A} \eta_X^P(x) = \phi$

This basic example is from treatment point of view rather uninteresting. However, it reveals the need for including treatment protocols, also based on medical conditions that must be integrated in the relation ρ^P. Adding levels of interactions brings this treatment a further step towards reality of drug treatment. Let us thus depart from the two- valued view of drug interactions and adopt the one described in Section 2. Excluding the possible interactions that are not tested, we denote the remaining lattice (total order) by L.

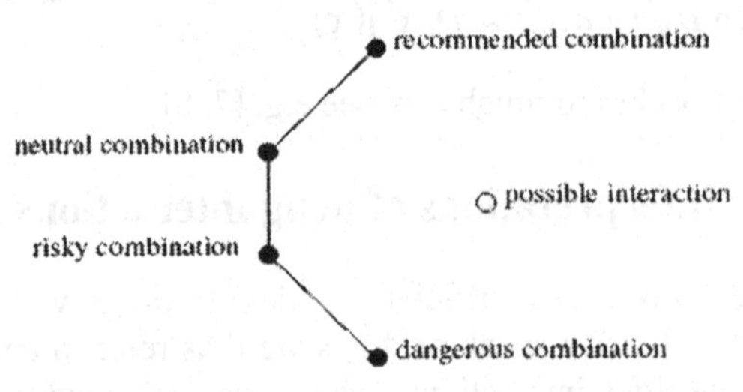

Fig. 1. Lattice representing levels of interaction

The interactions relation is now a mapping $\rho_X^L : X \to LX$, based on the many valued powerset monad (L, η, μ). Let further *M* be a set of medical conditions and let $\rho^L[M]$ be the subrelation of ρ which considers interactions with pharmacological treatments based on these medical conditions in *M*. A typical such medical condition is renal insufficiency in presence of which ACE inhibitors (C09A, C09AA) may have favorable effects. X

Interpretations can now be provided using rough monadic operators in the many-valued settings. By doing so we observe that the clinical usefulness of these interpretations comes down to defining $\rho^L[M]$ so as to correspond to real clinical situations. Operating with these sets then becomes the first step to identify connections to guidelines for pharmacological treatment. A full exploration of these details is beyond the scope of this paper.

Note also that *X might* not be the optimal choice of a universe. Mixing elements from various levels in the ATC hierarchy, and relating medical conditions both to therapeutic/pharmacological subgroups as well as to chemical subgroups, opens up a wide range of possibilities for rough monadic interpretations of interactions. In future papers we will explore these possibilities in full detail.

5 Conclusions and future work

Pharmacologic information is typically used for various dictionary purposes. National pharmacologic information systems have existed for decades and are now available in electronic form, some more elaborate, some involving rather straightforward presentations. Pharmacologic information used in connection with guidelines and decision support involving pharmacologic treatment are more challenging as they involve requirements to handle interactions, such as seen in pharmacologic treatment e.g. of public diseases.

Future work will be directed towards the interplay between interactivity of drugs and pharmacological treatment as described in international guidelines. Interactivity of drugs is a computational challenge involving uncertainties and various nearness aspects. Guideline implementations are challenging from logic point of view as guideline rules must be accurately represented and the underlying logic of a (set of) clinical guidelines must be well understood.

References

P. Eklund, M.A. Gal´an, *Monads can be rough*, S. Greco et al. (Eds.), Proceedings of The Fifth International Conference on Rough Sets and Current Trends in Computing (RSCTC 2006), Lecture Notes in Artificial Intelligence 4259 (Springer-Verlag Berlin Heidelberg 2006), pp. 77-84, 2006.

P. Eklund, M.A. Gal´an, W. G¨ahler, J. Medina, M. Ojeda Aciego, A. Valverde, *A note on partially ordered generalized terms*, Proc. of Fourth Conference of the European Society for Fuzzy Logic and Technology and Rencontres Francophones sur la Logique Floue et ses applications (Joint EUSFLAT-LFA 2005), 793-796.

P. Eklund, W. G¨ahler, *Partially ordered monads and powerset Kleene algebras*, Proc. 10th Information Processing and Management of Uncertainty in Knowledge Based Systems Conference (IPMU 2004).

W. G¨ahler, *General Topology – The monadic case, examples, applications*, Acta Math. Hun- gar. 88 (2000), 279-290.

W. G¨ahler, P. Eklund, *Extension structures and compactifications*, In: Categorical Methods in Algebra and Topology (CatMAT 2000), 181–205.

P. H´ajek, *Metamathematics of Fuzzy Logic*, Kluwer Academic Publishers, 1998.

J. J¨arvinen, *On the structure of rough approximations*, Fundamenta Informaticae 53 (2002), 135-153.

J. J¨arvinen, *Lattice Theory for Rough Sets*, Transactions on Rough Sets VI, Lecture Notes in Computer Science 4374 (Springer-Verlag Berlin Heidelberg 2007), pp. 400-498, 2007.

S. C. Kleene, *Representation of events in nerve nets and finite automata*, In: Automata Studies (Eds. C. E. Shannon, J. McCarthy), Princeton University Press, 1956, 3-41.

J. Kortelainen, *A Topological Approach to Fuzzy Sets*, Ph.D. Dissertation, Lappeenranta University of Technology, Acta Universitatis Lappeenrantaensis 90 (1999).

Kusiak, A., J. Kern, K. Kerstine, K. McLaughlin, and T. Tseng: 2000a, 'AutonomousDecision-Making: A Data Mining Approach'. *IEEE Transactions on Information Technol- ogy in Biomedicine* 4(4), 274–284.

Kusiak, A., K. Kerstine, J. Kern, K. McLaughlin, and T. Tseng: 2000b, 'Data Mining: Medical and Engineering Case Studies'. In: *Proceedings of the Industrial Engineering Research 2000 Conference*. pp. 1–7.

E. G. Manes, *Algebraic Theories*, Springer, 1976.

Z. Pawlak, *Rough sets*, Int. J. Computer and Information Sciences 5 (1982) 341-356. 15. M. Persson, J. Bohlin, P. Eklund, *Development and maintenance of guideline-based decision support for pharmacological treatment of hypertension*, Comp. Meth. Progr. Biomed., 61 (2000), 209-219.

J. Serra. *Image Analysis and Mathematical Morphology*, volume 1. Academic Press, 1982. A. Tarski, *On the calculus of relations*, J. Symbolic Logic 6 (1941), 65-106.

The sixth report of the joint national committee on prevention detection, evaluation, and treatment of high blood pressure, Technical Report 98-4080, National Institutes of Health, 1997.

Rough Set Approach to Analysis of Students Academic Performance in Web-based Learning Support System

Lisa Fan and Tomoko Matsuyama

Department of Computer Science, University of Regina, Regina,
SK, S4S 0A2, Canada
fan@cs.uregina.ca, tomoko.matsuyama@adxstudio.com

Abstract. This paper presents a Rough Set approach to analyze of students academic performance in a Web-based learning support system (WLSS). Web-based education has become a very important area of educational technology. This paper considers individual learners working alone without support from a teacher to provide guidance and advice on learning approach. Learners may have access to a wealth of material but may be faced with other problems such as material selection, planning a learning strategy, maintaining motivation and sequencing learning sub-goals. It might create a situation where some students may not be able to improve their grade as well as they could, compared to a face-to-face course. What if customized course materials were prepared for each student? It might fill this gap. Their records, such as grades for prerequisite courses or some personal factors that seems to affect their academic performance are used as student profile. In this paper, we discuss how to use Rough Sets to analyze student personal information to assist students with effective learning.
Decision rules are obtained using Rough Set based learning to predict academic performance, and it is used to determine a path for course delivery.

1 Introduction

Web based learning environments are becoming very popular. The web has become a powerful environment for distributing information and many educational providers are using it to deliver knowledge to an increasingly wide and diverse audience.
Typical web-based learning environments, such as Web-CT [1] include course content delivery tools, quiz module, virtual workspaces for sharing resources, grade reporting systems, assignment submission components, etc. They are powerful integrated systems that support a number of activities performed by teachers and students during the learning process [2]. Students who study a course on the Internet tend to be more heterogeneously distributed than those found in a traditional classroom situation. Therefore, the learning material should be presented in a more personalized way.

In a web-based learning environment, instructors provide resources such as text, multimedia and simulations, and discussions. Remote learners are encouraged to use the resources and participate in activities. However, it is difficult and time consuming for educators to thoroughly track and assess all the activities performed by the learners on these tools [3]. Moreover, it is hard to evaluate the structure of the course content and its effectiveness on the learning process. When instructors put together an on-line course, they may compile interactive course notes, simulations, examples, exercises, quizzes, web resources, etc. This on-line hyper-linked material could form a complex structure that is difficult for learners to navigate. Designers and instructors, when devising the on-line structure of the course and course material, have a navigation pattern in mind and assume all on-line learners would follow the predefined learning path; the path put forth in the design and implemented by some hyperlinks. Learners, however, could follow activities. Often this sequence is not the optimum learning sequence for individual learners, and does not satisfy the learner's individual learning needs [4].

Not all students have the same ability and skills to learn a subject. Students may have different background knowledge for a subject, which may affect their learning. Some students need more explanations than others. Other differences among students related to personal features such as age, interests, preferences, etc. may also affect their learning. Moreover, the results of each student's work during the learning session must be taken into account in order to select the next study topics to the student [5].

In this paper, we discuss how to use Rough Sets to analyze student personal information to assist students with effective learning. Decision rules are obtained using Rough Set based learning to predict academic performance, and it is used to determine a path for course delivery.

2 Related Work

It is difficulty to accurately measure the achievement of students. However such measurement is essential to any web-based learning support system. Liang et al [6] proposed a Rough Set based distance-learning algorithm to understand the students' ability to learn the material as presented. They analyzed students' grade information and form an information table to find rules behind the information table and use a table of results from course work to determine the rules associated with failure on the final exam. Then they can inform students in subsequent courses of the core sections of the course and provide guidance for online students. Magagula et al [7] did a comparative analysis of the academic performance of distance and on-campus learners. The variables used in their study are similar to ours. Lavin [8] gave a detailed theoretical analysis of the prediction of academic performance. Vasilakos et al [9] proposed a framework of apply computational intelligence - fuzzy systems, granular computing (including Rough Sets), and evolutionary computing to Web-based educational systems. Challenging issues such as knowledge representation, adaptive properties and learning abilities and structural developments must be dealt with.

3 Rough Set Overview

Rough Set is a mathematical tool to deal with vagueness and uncertainty [10]. It was introduced by Zdzislaw Pawlak in the early 1980s. The main problems that Rough Set deal with is data reduction, discovery of data dependency, estimate of data importance, and approximate classification of data etc. The major advantage of using Rough Set for data reduction and feature ranking is: Rough Set can generate many different reducts of a dataset at one time while other method can only generate one reduct for a dataset. A reduct is the minimal feature set having the same classification ability of the whole feature set. With more than one reduct, we can generate a dataset which is more stable than the dataset generated from only one reduct. At the same time, we can still reduce the redundant and noisy data from the original dataset.

In Rough Set theory, the reduct of knowledge is the most fundamental part of knowledge. It is a subset of the original attribute set, and can help us to reduce the dimensionality of datasets. By getting the reduct of the meta-attributes without losing information, we can make the recommendation procedure fast and efficient. Reduct can be demonstrate that **R** is a family of equivalence relations and let $R \in$ **R.** If IND(R)=IND (**R**-{R}), R is dispensable in **R**; otherwise R is indispensable in **R.** The family **R** is independent if each R∈**R** is indispensable in **R**. In the other word, it must satisfy the following two conditions: First, the original attribute set can be substituted by the reduct while preserving the accuracy of classification. Second, if any one of attributes is removed from the reduct, it will absolutely lead to new inconsistency.

4 Experiment Description and Settings

In this study, twelve variables were considered for the analysis based on an assumption that they seem to be related to academic performance. All of them are personal, such as individual background and study environment. It is not necessary to use very fine decimal values for finding out the likely hood of performance. Some of the values should be simple, descriptive and approximated so that analysis of the data and creation of the rules would be successful. Discretization of variables is needed, so that they are used in conjunction with other variables. Some of them might be replaced with binary value. In the rules to predict academic performance, discretization of the variables has very important role to show what they imply without any extra explanation.

The following is the discretization of variables:

AGE:	YOUNG, POST-YOUNG, PRE-MIDDLE, POST-MIDDLE, PRE-OLD, OLD
GENGER:	MALE, FEMALE
FINANCIAL:	SUPPORT, NOT SUPPORT
JOB:	YES, NO
ORIGIN:	ENGLISH COUNTRY, OTHER COUNTRY
ENVIRONMENT FOR STUDY:	GOOD, BAD
MARTIAL:	MARRIED, NOT MARIED
DEPENDENTS:	YES, NO

TIME FOR STUDY:	FAIR, NOT ENOUGH
MOTIVATION:	YES, NO
HEALTH:	GOOD, NO
MAJOR:	SCIENCE, ARTS
PERFORMANCE:	OVERANDEQUAL80, LESS80

4.1 Survey and Data Collection

A survey was made to collect data from the students on campus with the variables constructed previously. The sample size came out to twenty eight. In general, this size is not as big as other studies. However, this size of data is good to start with and there are varieties of age classes available for this study. As the process of constructing the survey proceeded, we realized that not only computer science knowledge, but also psychological and educational knowledge have important roles in order to collect accurate data for analysis. There are twelve variables prepared and considered for use in the analysis. Each variable has two or more choices that participants can select. A description of each choice should be understandable and concise. The discretization of variables was not performed on the variables for the survey. Simplicity is considered to be more important for the survey. For instance, instead of saying HAS_SUPPORT or DOES_NOT_SUPPORT, which is used in the database, choice would be YES or NO corresponding to a questionnaire. So that participants have less comprehension required. Positive choices, for instance, YES or GOOD should be prior to the negative description. This is based on the assumption that the majority of answer should be a positive choice rather than a negative one. If a negative response would be considered to be the majority, a questionnaire should be negated in order to have a majority of positive response. This gives participants greater ease to answer the survey. An explanation of each choice should only appear in the corresponding question. It reduces probability that the contents of the survey will be redundant. Considering these cares, questionnaires should not give participants any difficulty to comprehend the questions, because confusing them might affect the quality of the data, which might lead to constructing undependable rules.

For constructing data, discretization of variables has to be performed on some data in the survey. For instance, for AGE, the groups in the survey are described with numerical values. These are converted to descriptive groups, such as YOUNG, POST-YOUNG, PRE-MIDDLE, MIDDLE, POST-MIDDLE, PRE-OLD, OLD. The following table contains the data where descritaization is performed. The total size of the data is twenty-eight and the number of variables is thirteen.

ID	age	gender	finance	job	origin	houseEnvi	martial	dependents	time	moti	health	major	grade
1	young	male	support	yes	english	good	no	no	fair	no	good	arts	less80
2	young	male	support	no	english	good	no	no	fair	no	good	arts	less80
3	postmiddle	male	support	yes	english	good	no	no	fair	yes	good	arts	less80
4	young	male	support	yes	english	bad	no	no	fair	no	good	science	over&equal80
5	young	female	support	yes	english	good	no	no	fair	yes	good	arts	less80
6	young	male	support	yes	english	good	no	no	fair	no	good	arts	over&equal80
7	young	male	support	yes	english	good	no	no	fair	no	good	arts	less80
8	postyoung	female	support	yes	other	good	no	no	fair	yes	good	arts	over&equal80
9	young	male	support	no	english	bad	no	no	fair	yes	good	arts	over&equal80
10	young	male	no support	yes	english	good	no	no	fair	yes	good	science	over&equal80
11	postyoung	female	support	yes	english	good	no	no	fair	yes	good	science	over&equal80
12	young	female	support	yes	english	bad	no	no	fair	no	good	arts	over&equal80
13	young	male	support	yes	english	good	no	no	fair	no	good	science/ arts	less80
14	young	male	no support	no	english	good	no	no	fair	yes	good	science	over&equal80
15	postmiddle	male	no support	yes	english	good	yes	yes	fair	no	good	science	less80
16	middle	male	no support	yes	other	bad	yes	yes	not enou	yes	good	science	over&equal80
17	premiddle	male	support	yes	english	good	yes	yes	fair	yes	good	science	over&equal80

レコード: 14 | ◀ | 1 | ▶ | ▶| | ▶* / 28

Fig. 1 Data for analysis

Academic performance is a target attribute. The purpose of the analysis is to predict the degree of the students' performance, more than or equal to 80 % or less than 80 %. This degree could be changeable depending on what the purpose of the analysis is. For instance, if this outcome is used for regular prediction for future courses, all it wants to do is let students know whether or not they will have a learning curve or issues before it is too late. In this case, the scale might be around 70%. If the system would predict probability of a good performance, 80 % might be high enough.
These values form 2,048 classes that fall into either OVERANDEQUAL80 or LESS80. This does not mean that all of them have effect on the target attribute. Removing some of them does not change dependency. Instead of calculating probability and dependency manually, DQUEST was used to estimate dependency. The following is the reducts that used for analysis.

1.Gender, Job, Origin, Martial, Time, Motivation, Subject(16)
2.Age, Job, Environ, Time, Motivation, Subject(14)
3.Age, Gender, Job, Time, Motivation, Subject(19)
4.Gender, Financial, Job, Origin, Environment, Martial, Motivation, Subject(20)
5.Age, Financial, Job, Environ, Motivation, Subject(16)
6.Age, Gender, Financial, Job, Motivation, Subject(17)
7.Age, Gender, Financial, Job, Environment, Subject(17)

5 Results and Discussion

The numbers in parenthesis show the different combinations of these attributes appearing in the knowledge table. Low values represent strong data patterns. Since reduct #2: Age, Job, Environ, Time, Motivation, Subject (14) has the lowest value, we focused on the reduct for analysis.
Dependency is a measure of the relationship between condition attributes and the decision attribute. Significance is decreased in dependency caused by eliminating the attribute from the set. The following are rules by target attribute value. Cno implies the number of cases in the knowledge table matching the rule. Dno implies the total

number of cases in the knowledge table matching. Rules representing strong data patterns will have high values for both Cno and Dno.

Greater than or equal to 80 %	Cno, Dno
1.(Environ=bad)^(Time=fair)^(Subject=arts)	2,11
2.(Age=premiddle)^(Time=fair)	5,11
3.(Time=not)^(Subject=arts)	1,11
4.(Time=fair)^(Motivation=yes)^(Subjects=science)	5,11

Less than 80 %.	Cno,Dno
1.(Age=young)^(Motivation=no)^(Subject=science)	2,14
2.(Job=yes)^(Motivation=no)	1,14
3.(Age=young)^(Job=yes)^(Motivation=yes)^(Subject=arts)	3,14
4.(Age=postmiddle)	2,14
5.(Job=no)^(Time=no)	1,14
6.(Age=premiddle)^(Subject=arts)	1,14
7.(Time=no)	5,14

The rules are 2 and 4 for >=80% and 7 for <80%, since they have the highest value among the above rules.
(Age=premiddle)^(Time=fair)->(PERFORMANCE=OVERANDEQUAL80)(Time=fair)^(Motivation=yes)^(Subjects=science)->(PERFORMANCE=OVERANDEQUAL80)(Time=no) ->(PERFORMANCE= LESS80)

The rules imply that:
1. If one is in range of 30 ~ 35 years old and has enough time to study then academic performance most likely will be greater than or equal to 80%
2. If one has enough time, has motivation to study and majors in science, then academic performance most likely will be greater than or equal to 80%
3. If one does not have enough time then performance is less than 80%

So many factors influence student academic performance. Rough Set is used to analyze these factors in order to gain some knowledge. Using the knowledge and analysis of past grades allows us to predicting academic performance.

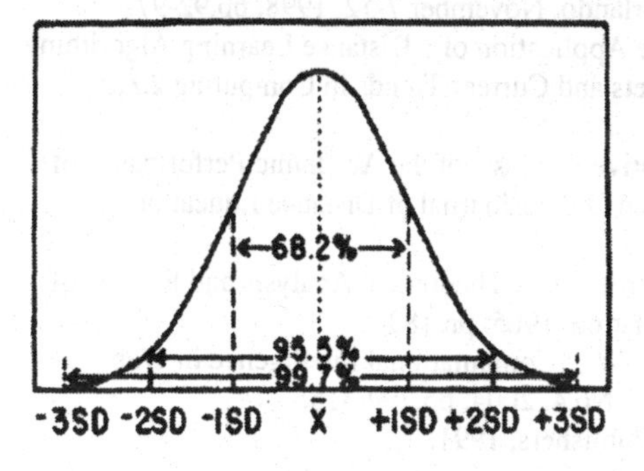

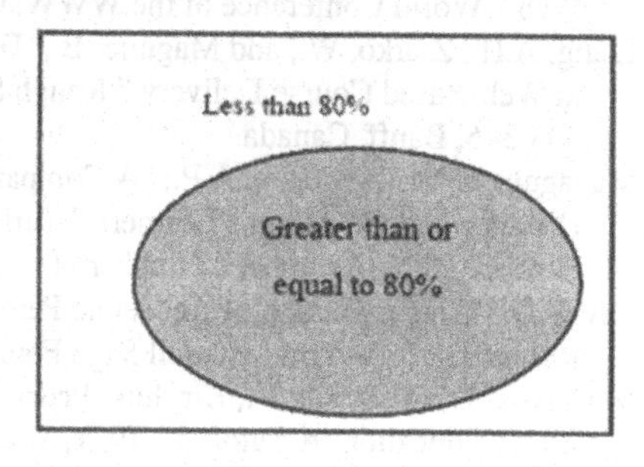

Fig. 2 visualization of standard deviation and rough set

6 Conclusion

In this paper, we have presented the Rough Set approach to analysis of student academic performance. The results will assist students in their learning effectively. It also guides the instructor to improve course contents.
Based on the results, the students can be divided into guided and non-guided groups. For the guided group, the course material would be sequentially presented with more teacher-centered contents, whereas the non-guided group, the course material is presented non-linearly. Students have more flexibility to explore the course material that is more student-centered. Figure 3 is the interface of the online course. Based on students learning styles, knowledge background, ability to learn, and personal features, Web-based learning support system provides a personalized learning environment to meet individual learner's need.

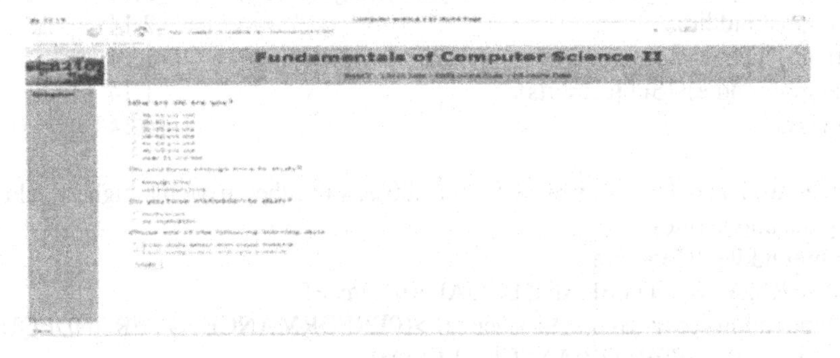

Fig. 3 Interface of the Web-based online course

References

WebCT, WebCT home page, http://www.webct.com/.

Brusilovsky, P. KnowledgeTree: A Distributed Architecture for Adaptive E-learning. WWW 2004, May 17-24, 2004, ACM 104-111.

Zaiane, O. R., "Building a Recommender Agent for e-Learning Systems", International Conference on Computers in Education (ICCE'02), Auckland, December 03-06, pp55-59.

Fan, L., "Personalization and Adaptation in Web-based Learning Support System" Proceedings of the Web-based Support Systems (WSS04), Beijing, China, 2004.

Brusilovsky, P., Anderson, J., "An adaptive System for Learning Cognitive Psychology on the Web", World Conference of the WWW, Orlando, November 7-12, 1998, pp.92-97.

Liang, A.H., Ziarko, W., and Maguire, B.,"The Application of a Distance Learning Algorithm in Web-Based Course Delivery." Rough Sets and Current Trends in Computing 2000: 338-345, Banff, Canada.

Magagula, C.M., Ngwenya, A.P., "A Comparative Analysis of the Academic Performance of Distance and On-campus Learners." Turkish Online Journal of Distance Education, October 2004 , Volume: 5 Number: 4

Lavin, D., "The Prediction of Academic Performance: A Theoretical Analysis and Review of Research." New York: Russell Sage Foundation, 1965. pp.182.

Vasilakos, T., Devedzic, V., Kinshuk, Pedrycz, W., "Computational Intelligence in Web-Based Education: A Tutorial", JILR, Vol.15,No.4, 2004, pp.299-318.

Pawlak, Z., "Rough Sets", Kluwer Academic Publishers, 1991.

Visual Data Mining of Symbolic Knowledge using Rough Sets, Virtual Reality and Fuzzy Techniques: An Application in Geophysical Prospecting

Julio J. Vald'es

National Research Council Canada Institute for Information Technology
M50, 1200 Montreal Rd.,Ottawa, ON K1A 0R6 julio.valdes@nrc-cnrc.gc.ca,
WWW home page: http://iit-iti.nrc-cnrc.gc.ca

Abstract. Visual data mining using nonlinear virtual reality spaces (VR) is ap- plied to symbolic knowledge in the form of production rules obtained by rough sets methods in a classification problem with partially defined and imprecise classes. In the context of a geophysical prospecting problem aiming at finding underground caves, a virtual reality nonlinear space for production rules is constructed. The distribution of the rough sets derived rules is characterized by a fuzzy model in both the original 5D space and in the 3D VR space. The membership function of the target class (the presence of a cave) is transferred from the rules to the data objects covered by the corresponding rules and mapped back to the original physical space. The fuzzy model built in the VR space predicted sites where new caves could be expected and one of them was confirmed.

1 Introduction

While applied frequently to databases, visualization techniques have not been applied often to the analysis of symbolic information. However, symbolic knowledge like for example, sets of production rules, are difficult to interpret for humans because of their more abstract nature and this is where visual methods become important aid. The purpose of this paper is to show that in addition to the understanding of symbolic knowledge provided by visual techniques, in particular virtual reality spaces [9], [11], mathematical models can be derived from the geometric properties of the symbolic objects in these spaces which can solve complex classification problems. In particular, situations where some of the classes are undefined because of lack of knowledge about class membership and where in addition, the classes themselves are *fuzzy*.

A general approach is proposed consisting of: i) use rough sets techniques for learning production rules from the original data using the imperfectly defined class labels ii) construct a nonlinear virtual reality space preserving the structure of the rules, iii) perform a data analysis in the new and the high dimensional space of the rules, iv) construct a fuzzy model based on the geometric properties of the rules in these spaces, v) induce the membership functions of the known classes to the database objects covered by the rules. This approach is applied to a real-world problem: the geophysical prospecting of caves, where class membership can be defined only for a certain subset of the database objects and where the classes (presence/absence of a cave) are fuzzy.

2 Virtual Reality Representation of Information Systems

Several reasons make Virtual Reality (VR) a suitable paradigm: it is is flexible, it allows immersion and creates a living experience. Of no less importance is the fact that in or- der to interact with a virtual world, no mathematical knowledge is required. A virtual reality based visual data mining technique, extending the concept of 3D modeling to information systems and relational structures, was introduced in [9], [11]. It is oriented to the understanding of large heterogeneous, incomplete and imprecise data, which includes symbolic knowledge. The objects are considered as tuples from a heterogeneous space $\hat{H}^n$ [10]. A virtual reality space is the tuple $\boldsymbol{r}$ =< $\underline{O}$; G; B; $\Re^m$; g_o; l; g_r; b; r >, where $\underline{O}$ is a relational structure ($\underline{O}$ =< O; Γ^v >, the O is a finite set of objects, and Γ^v is a set of relations), G is a non-empty set of geometries representing the different objects and relations. B is a non-empty set of behaviors of the objects in the virtual world. < is the set of real numbers and $\Re^m \subset$ $\mathbf{R}^m$ is a metric space of dimension m (Euclidean or not) which will be the actual virtual reality geometric space. The other elements are mappings: g_o: $O \rightarrow G$, φ: $O \rightarrow \Re^m$, g_r: $\Gamma^v \rightarrow G$, $b: O \rightarrow B$.

Several desiderata can be considered for building a VR-space [11]. From an unsupervised perspective, the role of ' could be to maximize some metric/non-metric structure preservation criteria (e.g. similarity) [2]. If ±ij is a dissimilarity measure between any two i; j∈ U (i; j∈ [1; N], where N is the number of objects), and $\xi_{i^v j^v}$ is another dissimilarity measure defined on objects i^v;$j^v \in$ O from $\boldsymbol{r}$ ($i^v = \xi(i)$; $j^v = \xi(j)$, they are in one-to-one correspondence). An error measure frequently used is [7]:

$$\text{Sammon error} = \frac{1}{\sum_{i<j} \delta_{ij}} \frac{\sum_{i<j} (\delta_{ij} - \xi_{i^v,j^v})^2}{\delta_{ij}} \tag{1}$$

3 The Data Mining Process

The original data is processed with Rough Sets techniques and rules are obtained relating the prediction attributes with the classes (this result is considered partial; if not all of the class are known). Then a virtual reality space for the obtained rules is built (using the method of Fletcher-Reeves [5]) and an analysis of the rules in the original and in the new spaces is made. Finally, the results of the analysis are mapped back into the original physical space for interpretation. Fig.-1.

3.1 Application to a Geophysical Prospecting Problem

Cave detection is a very important problem in civil and geological engineering. Typically caves are not opened to the surface and geophysical methods are required for their detection, which is a complex task. In a pilot investigation, geophysical methods and a topographic survey were used with the goal of deriving criteria for predicting the presence of underground caves [8]. In the studied area, a cave was known to exist, but the presence of others was suspected.

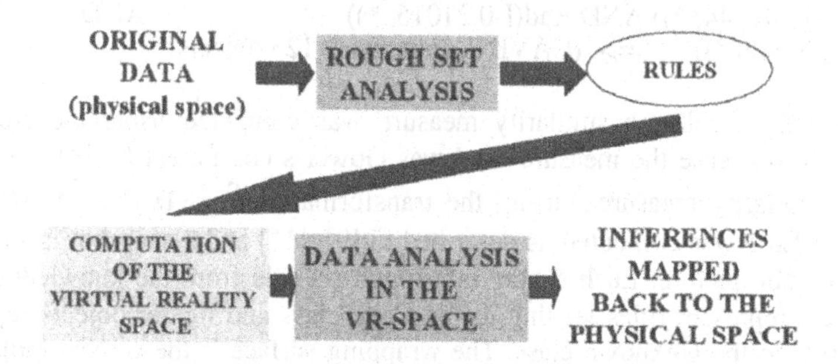

Fig. 1: The data analysis process.

This is a problem with partially defined classes: the existence of a cave beneath a measurement station is either known for sure or unknown (i.e. only one class membership is really defined). Moreover, the classes themselves are also imprecise or *fuzzy*, as there are no sharp boundaries between the classes. The problem is not the typical two-class presence/absence one because only one class is known with certainty: a combination of unsupervised and supervised approaches is required.
Five geophysical methods were measured on a regular grid containing 1225 measurement stations (the objects) [8]. The measured fields were: i) the spontaneous electric potential (SP_{dry}) at the earth's surface measured during the dry season, ii) the vertical component of the electro- magnetic field in the VLF region of the spectrum (frequency range [3 ¡ 30] kHz), iii) the spontaneous electric potential measured during the rainy season (SP_{dry}), iv) the gamma ray intensity (*Rad*) and v) the topography (*Alt*). A data preprocessing process was performed consisting of: i) conversion of each physical field to standard scores (zero mean and unit variance), ii) model each physical field f as composed of a trend, a signal and additive noise: f(x,y) = t(x,y) + s(x,y) + n(x,y) where t is the trend, *s* is the signal, and n is the noise component, iii) fitting a least squares two-dimensional linear trend

$\hat{t}(x,y)=c_0+c_1x+c_2y$ and computation of the residual: $\hat{r}(x,y) = f(x,y)- \hat{t}(x,y)$, iv) Convolution of the residual with a low-pass zero-phase shift two-dimensional digital filter [3] to attenuate the noise component, and v) Re-computation of the standard scores and addition of a class attribute indicating whether a cave is known to exist below the corresponding measurement station or if it is unknown. The pre-processed data set will be called prp-data.

Rough set analysis was performed using the Rosetta system [6]. The prp-data was discretized using the Boolean Reasoning algorithm and reducts were computed. Only one reduct was found containing all of the five attributes, indicating that none of the observed geophysical fields can be discarded without loosing discernibility. A set of 345 rules were obtained from the reduct and the following are two examples:

SPdry([*, -1.50209)) AND VLF([*, -1.14882)) AND
SPrain([*, -0.46789)) AND Rad([*, -1.54413)) AND
Alt([*, -1.22398)) => (CAVE is present) (6 objects)

SPdry([-0.16981, *)) AND VLF([-0.75462, *)) AND
SPrain([0.48744, *)) AND Rad([-0.21015, *)) AND
Alt([0.00346, *)) => (CAVE is unknown) (123 objects)

For each pair of rules, a similarity measure was computed using the condition attributes. In this case the measure used was Gower's coefficient (s) [4], converted into a dissimilarity measure δ using the transformation $\delta = (1/s) -1$. A VR-space minimizing Eq.1 was computed as described in [9], [11] and a snapshot is shown in Fig.2 as a static picture. Each sphere is a rough set rule from the knowledge base: dark objects represent rules leading to the Cave class and lighter objects represent rules leading to the Unknown class. The wrapping surface is the convex hull of the Cave class wrapping all of its rules (computed according to [1]) and the star indicates its centroid. There are rules leading to the unknown class which are within the hull of the cave class, indicating that they are similar to those concluding about the presence of a cave. Another subset of the rules concluding about the unknown class is located outside of the surface enclosing the set of rules of the cave class. They are more representative of the no-cave situation. In Fig.2, the distance d between any rule in the space and the centroid of the Cave class is shown.

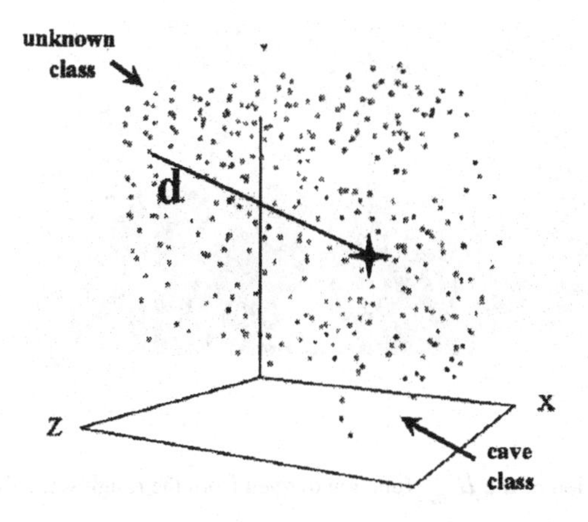

Fig. 2: Snapshot of the VR space containing the rules obtained via Rough Sets. Dark objects: cave class. Light objects: unknown class. Many objects of the unknown class are within the cave class.

The distance between any rule in the space and this centroid gives an indication about how similar the corresponding rule is of being a descriptor of the cave properties as an abstract concept. This notion can be formalized as a fuzzy property with a membership function constructed, among many others, as:

$$\mu_{cave}(r_i) = 1 - \frac{d(r_i, c)}{d_{\max}} \quad (2)$$

where μ_{cave} (ri) is the membership of rule r_i to the cave class, $d(r_i,c)$ is the distance between the i-th rule and the centroid c of the cave class and dmax is the maximal distance between the centroid and the farthest rule.

Two μ_{cave} functions were computed: i) in the original 5-D space of the attributes appearing in the condition part of the rules and ii) in the VR 3-D space. Since rules are abstract symbolic entities (without any physical location), these results have to be mapped back to the physical space. This was done by transferring the membership to the cave class from each given rule to the objects covered by the rule, which correspond to the measurement points on the earth's surface (the physical space). Thus, each fuzzy membership function μ_{cave} (ri) of the rules leads to a two dimensional fuzzy member- ship function of the objects with respect to the cave class $\mu^{o_j}_{cave}$ (x,y), where (x,y) are the coordinates corresponding to the j-th data object oj, covered by rule ri. The distribution of the fuzzy memberships computed in the original 5-D space and in the 3-D VR-spaces are shown in Fig.3.1 (left and right respectively), as well as the map of the area, with the location of the known cave (Fig.3.1-center).

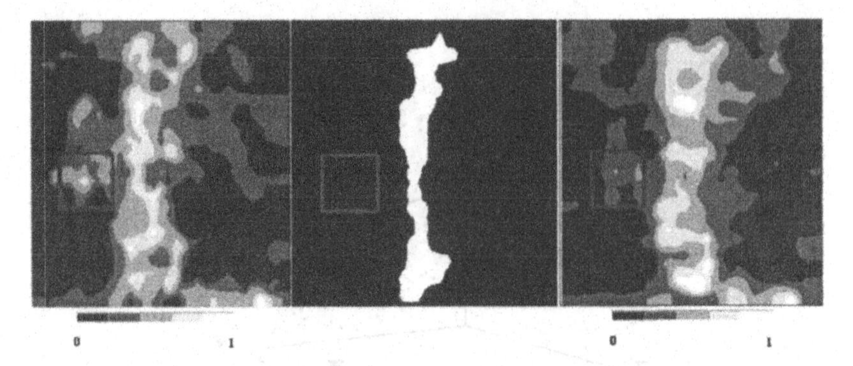

Fig. 3: Spatial distribution of the μ_{cave} function derived from the rough sets rules. Left: μ_{cave} from the original 5-D space. Center: The area with the known cave. Right: μ_{cave} from the VR space. In all images, a square shows the area where a borehole hit a previously unknown cave.

The fuzzy membership function in the original 5D space (Fig.3.1-left) has a central narrow band of high values which corresponds to the location of the known cave. In addition, there are other areas of high values located at the center-left and bottom - right, both beyond the outline of the surveyed cave. This suggests the presence of other caves, not opened to the surface. These areas are wrapped by a medium membership value enclosure emerging from the one enclosing the known cave which suggests that they might be a part of the same cave system.
A similar behavior is exhibited by the fuzzy membership function in the nonlinear VR space, shown in Fig.3.1 bottom-left. The patterns observed are the same in terms of the appearance of a central band of high values and the two additional areas of high membership values. The results can be perceived as more clear because the function is smoother. This indicates that the information lost during the nonlinear mapping of the original 5D space to the 3D space actually increased the signal to noise ratio, which is a very important feature. Some time after the geophysical investigation was made; a borehole was drilled in the location corresponding to the center-left area of high membership indicated above. A a cave was hit, thus confirming the results suggested by the presented approach.

4 Conclusions

Visual data mining of symbolic knowledge obtained with rough sets proved to be effective in understanding complex problems with partially defined and imprecise classes. Fuzzy models derived from the original rough set rules and from a virtual reality space obtained from them by nonlinear mapping, revealed the essential properties of the target class. In the studied case of a geophysical prospecting problem, it allowed the identification of areas where the presence of new hidden caves could be expected and one was confirmed by drilling. The comparisson of the fuzzy membership function in the original and in the VR space turned out to be a very effective noise reduction filter, which also preserves most of the information associated with the target class. This approach is domain-independent and could be applied to similar problems in other areas.

References

Barber, C. B., Dobkin, D. P., Huhdanpaa, H. T.: The Quickhull algorithm for convex hulls. ACM Trans. on Mathematical Software, 22(4),(1996). pp469–483.

Borg, I., and Lingoes, J., Multidimensional similarity structure analysis: Springer-Verlag, (1987), 390 p.

D.E. Dudgeon, R.M. Mersereau. Multidimensional Signal Processing. Prentice Hall, 1984. 4. Gower, J.C., A general coefficient of similarity and some of its properties: Biometrics, v.1, no. 27, p. 857–871. (1973).

Press, W.H.,Teukolsky S.A., Vetterling W.T., Flannery B.P: Numerical Recipes in C. The Art of Scientific Computing. Cambridge Univ. Press, 994 p. (1992).

Øhrn A., Komorowski J.: Rosetta- A Rough Set Toolkit for the Analysis of Data. Proc. of Third Int. Join Conf. on Information Sciences (JCIS97), Durham, USA, (1997), 403–407.

Sammon, J.W. A non-linear mapping for data structure analysis. IEEE Trans. on Computers C18, p 401–409 (1969).

Vald´es J.J, Gil J. L. Joint use of geophysical and geomathematical methods in the study of experimental karst areas. *XXVII International Geological Congress*, pp 214, Moscow, 1984.

Vald´es, J.J.: Virtual Reality Representation of Relational Systems and Decision Rules: An exploratory Tool for understanding Data Structure. In Theory and Application of Relational Structures as Knowledge Instruments. Meeting of the COST Action 274 (P. Hajek. Ed). Prague, November 14–16, (2002).

Vald´es, J.J : Similarity-Based Heterogeneous Neurons in the Context of General Observational Models. Neural Network World. Vol 12, No. 5, pp 499–508, (2002).

Vald´es, J.J.: Virtual Reality Representation of Information Systems and Decision Rules: An Exploratory Tool for Understanding Data and Knowledge. Lecture Notes in Artificial Intelligence LNAI 2639, pp. 615-618. Springer-Verlag (2003).

References

Barber C.B., Dobkin D.P., Huhdanpaa H.T.: The Quickhull algorithm for convex hulls. ACM Trans. on Mathematical Software, 22(4) (1996) pp.469–483.

Borg I. and Lingoes J.: Multidimensional similarity structure analysis. Springer-Verlag (1987) 390 p.

D.E. Dudgeon, R.M. Mersereau: Multidimensional Signal Processing. Prentice Hall, 1984. 4.

Gower J.C.: A general coefficient of similarity and some of its properties. Biometrics, Vol. 1 no. 27, p. 857–871 (1973).

Press W.H., Teukolsky S.A., Vetterling W.T., Flannery B.P.: Numerical Recipes in C. The Art of Scientific Computing. Cambridge Univ. Press, 994 p. (1992).

Øhrn A., Komorowski J.: Rosetta: A Rough Set Toolkit for the Analysis of Data. Proc. of Third Int. Join Conf. on Information Sciences (JCIS97), Durham, USA, (1997), 403–407.

Sammon J.W.: A non-linear mapping for data structure analysis. IEEE Trans. on Computers C18, p 401–409 (1969).

Valdés J.J., Gil J.L.: Joint use of geophysical and geomathematical methods in the study of experimental karst areas. XXVII International Geological Congress, pp 214, Moscow, 1984.

Valdés J.J.: Virtual Reality Representation of Relational Systems and Decision Rules: An exploratory Tool for understanding Data Structure. In Theory and Application of Relational Structures as Knowledge Instruments. Meeting of the COST Action 274 (P. Hajek, Ed). Prague, November 14–16, (2002).

Valdés J.J.: Similarity-Based Heterogeneous Neurons in the Context of General Observational Models. Neural Network World. Vol 12, No. 5, pp 499–508, (2002).

Valdés J.J.: Virtual Reality Representation of Information Systems and Decision Rules: An Exploratory Tool for Understanding Data and Knowledge. Lecture Notes in Artificial Intelligence LNAI 2639, pp. 615–618. Springer-Verlag (2003).

Section 4

International Workshop on Ubiquitous and Collaborative Computing

A Multi-Stage Negotiation Mechanism with Service-Oriented Implementation

Jiaxing Li[1] ,Chen-Fang Tsai[2],Yinsheng Li1 Jen-Hsiang Chen[3], Kuo-Ming Chao[1]

1 Fudan University, Shanghai, Software School, P.R.China,
{liys, 042053003}@fudan.edu.cn;
2 Aletheia University, Department of Industry Management Taiwan, tsai@email.au.edu.tw
3 Shih Chien University, Department of Information Management,Taiwan,
jhchen@mail.kh.usc.edu.tw

Abstract. Negotiation is an important aspect of collaborating and conducting businesses where multi-party interaction is involved. For the last few years, an automated negotiation has attracted attention from both research communities and industry. An automated negotiation mechanism can be built in a computing environment by simulating human negotiation process. This paper proposes a multiple-stage bi-lateral automated negotiation mechanism, including a co-evolutionary negotiation strategy, a sophisticated negotiation protocol and a game-theoretic negotiation method which is utilized to distribute the payoffs. The proposed mechanism is intended to address the search for joint efficiency of all involved negotiating parties in large and complex problem spaces. Using this structure it is possible to refine and explore potential agreements through an iterated process. The proposed automated negotiation approach is based on a specific game theory method and a heuristic approach.

1 Introduction

Negotiation is an important aspect of collaborating and conducting businesses where multi-party interaction is involved. Negotiation mechanisms are built in computer systems by simulating human reasoning mechanism. Negotiation has also been envisioned as a key area to improve the efficiency of E-businesses.
With the number of companies using E-business to support their business operations showing a dramatic increase, E-business is playing a more important role for enterprises. However, current E-business technology lacks flexibility and is limited to few scopes, especially for negotiation over business issues such as product prices, quantities, quality, and delivery time [1]. Therefore, it is desirable to have an automated negotiation mechanism in E-business that can improve the efficiency of e-business, as well as the quality of service, and the accuracy of complex cost calculations.

We have proposed in our previous work [2, 3, 4] an integrated negotiation mechanism which combines a co-evolutionary method with a game theory approach. The proposed approach can locate a solution, which is Pareto efficient and is close to a Nash equilibrium. This approach has led to some contributions to agent-based automated negotiation and has been applied to different problem domains. The paper is structured as follows. Section 2 describes related works and objectives of automated negotiation. Section 3 presents the proposed automated negotiation approach, which is based on a specific game theory method and heuristic approaches such as GA. Section 4 describes the implementation of the proposed automated negotiation approach from a service-oriented perspective, and gives a brief case study. Section 5 examines the experimental results and concludes the paper.

2 Objectives and Related Work

As a field of research, automated negotiation has attracted a number of academic groups. Among them, there is particular interest in the simulation of human negotiation in a computing environment. Agent technologies have been introduced to enable agents, acting as delegates for users, to conduct negotiations over conflicting issues between participating parties. This research aims at improving existing automated negotiation techniques so as to facilitate practical negotiation in E-business. Various theories and methods are proposed. These include: the argumentation approach [5, 6], game theory [7, 8], and Genetic Algorithms (GA) [1, 9, 10, 11]. All these methods have their strengths and weaknesses. To improve the efficiency and effectiveness of negotiation, there are several points which this research attempts to address:

2.1 The Quality of the Agreement

The quality of an agreement can be assessed in both an objective and a subjective way. In the objective context two important concepts arise: the idea of Pareto efficiency and the idea of equilibrium [12]. Pareto efficiency occurs if there is no other allocation of utility which can improve one negotiator's return without detriment to another negotiator. In the two negotiator situation the objective view can be shown on a diagram plotting the returns to each negotiator as the two axes. The idea of an equilibrium position is that any negotiator cannot improve his return unless the others make a change in their position. The objective view allows equilibrium positions to be verified. Ideally the result of negotiation is an equilibrium agreement that is Pareto efficient.

The subjective view of the value of the agreement will differ from the objective view if there is incomplete information. Each negotiator will be only aware of the payoffs to him. In these cases it will not be possible for any individual negotiator to know if a given agreement is Pareto efficient nor to confirm its objective equilibrium status. As a result it is likely that sub-optimal agreements will be reached. Without providing complete information, a negotiation process that allows partial information about preferences to be exchanged may make it possible to approach an optimal agreement.

2.2 The Negotiation Process

The context in which a negotiation takes place will determine the options available for the negotiators. The channels of communication and the allowed protocols will provide constraints on the targets and the strategies that can be adopted. The context must support sufficient exchanges of information to provide for an improvement process. It must be possible to progress towards an optimal agreement.
The individual concern in a negotiation will be the choice of strategy within the agreed context. One view of the choice of strategy is that the aim is to resolve the conflict between maximising private payoffs and enlarging the common benefit. Negotiators may give appropriate concessions that may increase common marginal utility. Negotiators have their own private strategies to analyse the current situation. Assuming a context in which alternating offer and counter offers are made the strategy must determine those offers and counter offers generated during the bargaining process. The problem is to design a strategy suitable for automation that will move towards an optimal solution.

2.3. Utility in a Multi-Issue Situation

Some types of negotiation work issue-by-issue, so the utility function can be computed independently for individual issues. However, in some cases a utility function needs to manage a set of interdependent issues. In a manufacturing supply chain both negotiators frequently negotiate over the three issues: price, quantity, and delivery date, trading one against the others. It is also true that the costs and/or benefits associated with different levels for these negotiated variables can involve a wide range of other internally relevant variables. This type of negotiation has to consider all these variables in its negotiation strategy. This will not be possible in a simple issue-by-issue form. However, some more complex form of issue-by-issue negotiation may be possible. For this reason, careful design of the strategy and its contextual protocol is required.

3 A Multi-Stage Negotiation Mechanism

The proposed automated negotiation approach is based on a specific game theory method and a heuristic approach. The advantage of a heuristic learning mechanism is that it provides flexibility and allows improvement in imprecisely understood contexts. Where negotiation is in a dynamic environment, actions are non-deterministic, in the sense that they do not always lead to the same results, when executed in the same state. The range of mechanisms applies different approaches to process simulation and the evaluation of the set of offer and counter-offer. The main differences in the approaches of the three kinds of mechanism are in the encoding and definition of objective functions. This research employs GA to produce an optimized solution over a large and complex search space through the evolutionary process.
In this research the GA approach is used in two ways. Initially, the possible solutions and/or problems are encoded in the chromosomes (e.g. the multi-issue is encoded into one chromosome). The system randomly generates a set of possible solutions to the problems that form a population. The GA method is then employed as part of the selection of strategies. In addition the calculation of the inverse of the function mapping a proposed offer to a utility value is also carried out using a GA.

A weakness of the heuristic approach is that it may lead to a sub-optimal solution as it may not examine the full space of possible outcomes and adopt only an approximate notion of rationality. One possible way to overcome this natural weakness is to improve results through a game theory related mechanism that moves towards Pareto optimal solutions. This research uses a game theory negotiation approach, the Trusted Third Party (TTP), to overcome the heuristic approach's natural weakness.

The TTP mediated game is extended from the traditional game by incorporating a negotiation mechanism. In the game, agents do not only reason on the game matrix but also attempt to find the equilibrium using given negotiation mechanisms. This approach is based on an assumption that the payoffs in the matrix are tradable. The advantage of this mechanism is to solve problems in difficult games where one negotiating agent does not know the other agent's payoffs (difficult games are those that have no Nash equilibrium or multiple Nash equilibrium).

TTP includes two communication actions and the corresponding negotiation protocol. Two communication actions provide a way to trade payoffs and therefore change the game matrix from a difficult game into an acceptable one. The general negotiation protocol is a loop of making or accepting a suggestion or making a counter suggestion.

3.1 The Proposed Approach for Automated Negotiation

In order to facilitate effective automated negotiation, we propose a multiple-stage negotiation architecture to support the negotiation process. Agents are enabled to generate and select effective strategies that lead to high payoffs. The aim of the proposed architecture is to enable negotiating agents to produce an agreement that is in equilibrium at their best joint efficiency position.

In the proposed negotiation mechanism, each stage includes two processes: a co-evolution process and a game theoretical reasoning process. The purpose of the co-evolution process is for the agents to identify the most appropriate negotiation strategies according to their personal gains and the joint efficiency. The selected negotiation strategies will be used by the agents to play against each other to form a payoff matrix. The resulting matrix will be reasoned with using the designated game theoretical approach, the Trusted Third Party (TTP) Mediation Game, in order to obtain a possible agreement for both agents. The agreed point, if there is one, should comply with the solution concepts of Pareto Efficiency and Equilibrium. The agreed deal will become a reference point for the agents to refine their search space by replacing their original expected gained utilities with the newly obtained potential deal. The second or later stages of negotiation repeat the same procedures in an attempt to gain a better deal for both agents. The stages will iterate until the negotiation ends. The termination of a negotiation process occurs when the predefined deadline approached or the agents agree to stop the process.

In the co-evolution process, each agent includes two Genetic Algorithms (GAs): Issue GA and Strategy GA. The Issue GA is driven by a selected negotiation strategy generated by the Strategy GA to locate offers with appropriate values for multiple negotiation issues. Therefore, a negotiation strategy contains a sequence of target utilities. The desired gained utility formulated by the negotiation strategy feeds into the Issue GA to carry out a search. The contents of the offers are, in fact, generated by the Issue GA. This is to prevent the agent from making irrational decisions or offers given that the negotiation issues have intricate non-linear interdependencies which are specified in the utility functions.

The processing time of the co-evolution process depends on the complexity of the utility functions and the parameter settings in the Issue GA and the Strategy GA. In order to facilitate the agents to shorten the search process, a concept 'degree of satisfaction' is introduced to the agents when they exchange offers. When an agent rejects an offer, it does not only make a counter offer, but also expresses its degree of satisfaction with the offer it received. So, an agent can take this information into account to generate the next offer. The degree of satisfaction is measured by the ratio of the agent's actual gain, if the agent accepts the offer, to the expected maximum. If the agents have similar bargaining powers, the appropriateness of the selected negotiation strategies can be measured by the difference between the agents' degrees of satisfaction with the deal. A small difference of degrees of satisfaction is better than a large one.

Joint efficiency is another measurement that can be used to evaluate a pair of negotiation strategies when the agents take equal weightings for increasing their own satisfaction and increasing their opponent's satisfaction. In this case, the agents will select the strategies which can produce a larger sum for the degrees of satisfaction.

This architecture involves numerous parameters (e.g. parameters settings for the GAs etc) and the selection of the values for them could have a significant impact on the negotiation behaviours and the results.

4 Service-Oriented Implementation and Case Study

SOA (Service-Oriented Architecture) is becoming a leading paradigm for information planning and application integration. Web services and Semantic Web are emerging as promising technologies to promote service-oriented architecture and service-oriented e-business systems. Therefore, this research designed and implemented the proposed negotiation mechanism from a service-oriented perspective, with the negotiation agent exposed as negotiation services providing open and standard application interfaces. Also, in order to apply this proposed negotiation mechanism in real business scenarios, a multi-model driven development platform has been developed so that the negotiation mechanism can be easily-modelled and integrated with other business processes.

As shown in Figure 1, the platform has three views for enterprises to plan and develop e-Business solutions, i.e., a business view, a process view, and a Web services view. To keep consistent among the three views, a set of rules have been devised to transform the objects among the three views automatically. To create an e-business solution, the enterprise starts with the business view and identifies involved business services, IT services, service providers and their relationships for the given business. Among them can be a negotiation step or service, where a set of business items need to be negotiated between participating parties. Then, the business properties of the selected processes can be defined. At last, a collection of negotiation parameters can be set in the service view including service metadata, strategy GA parameters, issue GA parameters and other negotiation parameters.

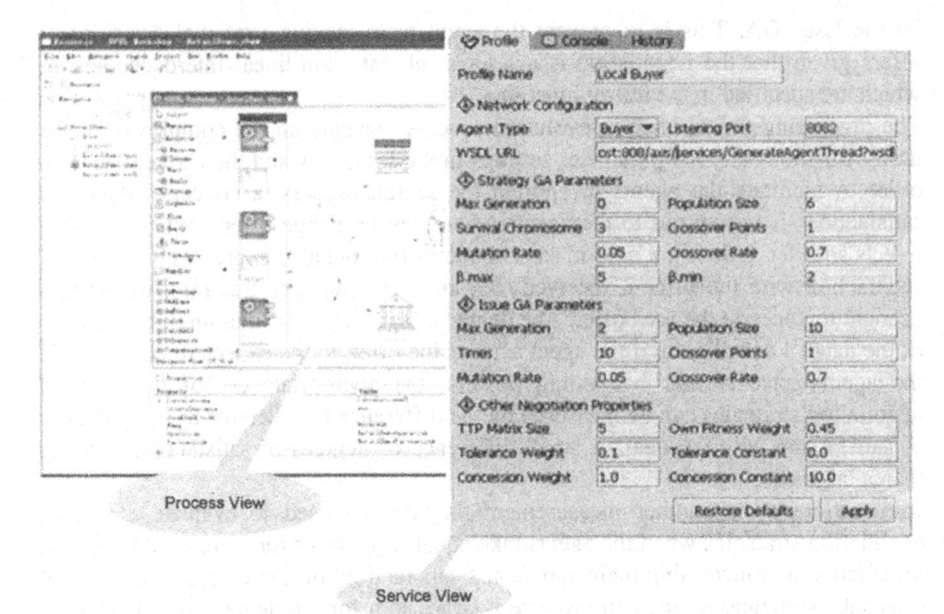

Fig. 1. Multi-model driven collaborative platform for automated negotiation (Service View)

In a typical retail supply chain, there are consumers, retailers (stores, distribution centers, and logistics), transportation providers (carriers), credit authorities, and possibly independent warehouse operators. There is an inventory tracking service to check shelves and inventory in the retail stores. Once the products are under a specified point, they request the distribution center to check the repository. Shipment and transportation are arranged if they have requested items at repository. The retailers' distribution center can choose the third-party transportation providers and warehouse operators. If the requested items are out or insufficient, the distribution center requests the logistics to purchase. The logistics or headquarter checks the partner suppliers, requests prices and their credits and then they start to negotiate with selected supplier on the price, amount, delivery time, payment and quality etc. Once an agreement is reached, the headquarter places orders to the suppliers. The suppliers will accept quotation and reply with rates.

In the above case, we demonstrate our proposed architecture with a buyer and a supplier scenario. In this scenario, there are two participating parties, a buyer who is a distributor and a supplier who is a manufacturer, carrying out negotiation on a potential business transaction. For both agents the three issues are price, quantity and delivery. The utility is measured by profit. The agents calculate the profits from given values associated with these issues. Figure 2 shows the payoff derived from the second stage negotiation which satisfies the concept of Pareto-efficiency and the second stage negotiation does improve the outcomes for both agents.

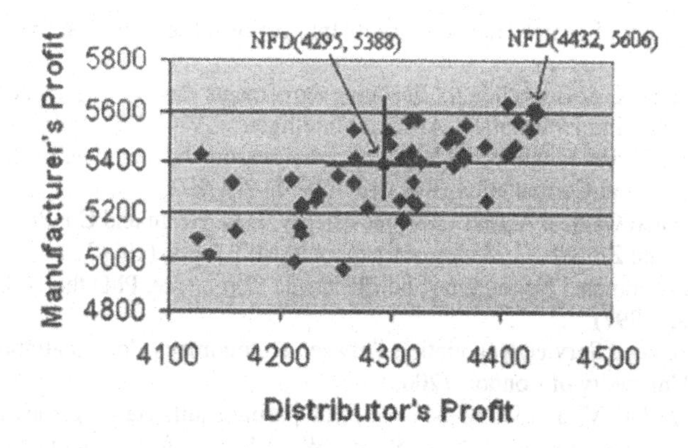

Fig. 2. Payoff Distribution in Two Stage Negotiation (NFD, an equilibrium, is derived from the TTP game. The total utilities for manufacturer and distributor are shown in the right and left respectively)

5 Conclusions

In this work, we have reported an automated negotiation mechanism which features evolution of negotiation strategy and an effective multiple-stage negotiation protocol. In the proposed system, in order to search complex and large spaces to find joint efficiency between negotiating parties, a co-evolutionary mechanism is employed to evolve negotiation strategies. Further, a TTP game theoretical approach is introduced to distribute the payoffs generated from the co-evolutionary approach so that we can overcome the heuristic approach's natural weakness.

The proposed multiple-stage negotiation protocol allows the equilibrium (the output of co-evolution) as input that feeds into the co-evolutionary approach to carry out further exploration of space and refinement of negotiation strategies in order to find a better solution.

The design of the proposed negotiation mechanism is taken into consideration from a service-oriented perspective, with the negotiation agent exposed as negotiation services with open and standard application interfaces. Also, in order to apply this proposed negotiation mechanism in real business scenarios, a multi-model driven development platform has been developed so that the negotiation mechanism can be easily-modelled and integrated with other business processes.

References

Matos, N., Sierra, C. and Jennings, N. R., Determining successful negotiation strategies: an evolutionary approach Proc. 3rd Int. Conf. on Multi-Agent Systems, Paris, France, (1998) 182-189.

Chao, K-M, Anane, R., Chen, J-H, and Gatward, R., Negotiating Agents in a Market-Oriented Grid, 2nd IEEE/ACM International Symposium on Cluster Computing and the Grid, IEEE CS, (2002) 436-437.

Chen, J-H, R. Anane, Chao, K-M, and Godwin, N., Architecture of an agent-based negotiation mechanism, Proceedings of The 22nd IEEE International Conference on Distributed Computing Systems Workshops, IEEE CS, (2002) 379-384.

Chen, J.H., Chao, K-M, Godwin, N., Reeves. C., and Smith, P., An automated negotiation mechanism based on co-Evolution and game theory, the 17th ACM Symposium on applied Computing, Special Track on Agents Interactions, Mobility, and System, (2002) 63-65.

Kraus, S., K. Sycara, K. & Evenchik A., Reaching agreements through argumentation: a logical model and implementation, Artificial Intelligence, Vol. 104 (1998) 1–69

Parsons, S., Sierra, C. and Jennings, N. R., Agents that reason and negotiate by arguing, Journal of Logic and Computation, Vol. 8(3), (1998) 261-292.

Binmore, K., Fun and Games: A Text on Game Theory, D.C. Heath and Co. Press, (1992).

Rosenschein, J.S. and Zlotkin, G., Rules of Encounter, MIT Press, (1994)

Oliver, J. R., On Automated Negotiation and Electronic Commerce, PhD thesis, University of Pennsylvania, (1997)

Peyman, F., Automated Service Negotiation Between Autonomous Computational Agents, PhD thesis, University of London, (2000)

Fatima. S, Wooldridge. M. and Jennings. N. R., Comparing equilibria for game theoretic and evolutionary bargaining models, Proc. 5th Int. Workshop on Agent-Mediated E-Commerce, Melbourne, Australia, (2003) 70-77

Muthoo, A., Bargaining and Applications. Cambridge University Press, (1999)

Activities Distribution Around a Horizontal Interface Supporting Small Group Meetings

Chiara Leonardi, Massimo Zancanaro, Fabio Pianesi

Bruno Kessler Foundation -IRST, Trento, Italy {cleonardi;zancana;pianesi}@itc.it

Abstract. In this paper, we present a preliminary field study aimed at investigating collaborative work activities supported by a horizontal interface. We analyzed the use of system by focusing on activities distribution and on coordination dynamics during actual meetings. The results have implications for the understanding of social dynamics occurring when groups of users explore horizontal displays.

1 Introduction

The new line of research in the fields of tabletop interaction and large-display groupware is trying to address the challenge of supporting human-human interaction beyond desktop-metaphor-based computing. Tabletop devices and large-displays–both vertical and horizontal—have demonstrated their value in supporting a range of work activities and tasks. Yet, recent studies also pointed out the challenges raised by these systems as well as the failures they are exposed to [O'Hara, Perry, Churchill and Russel 2004]. Besides the well-known problems of desktop groupware [Grudin 1994], these collaborative collocated applications present new problematic aspects [Huang, Mynatt, Russel and Sue 2006]. The public nature of the surface can increase uncertainty because of the lack of knowledge about what people have accessed, or who is controlling what; tensions can emerge when the display contravenes the norms of social acceptability and physicality, for example being forced to stand close, side-by-side in front of a vertical display can cause awkwardness and the feeling of invading each other's personal spaces [Rogers and Rodden 2004]. Furthermore, public displays can fall in disuse because potential users are less prone to exploration than to actual use of a system. Users maybe however be encouraged to discover helpful uses of the system by observing how other users interact with the system [Huang et al. 2006].

In the design of our system, called Collaborative Workspace, we tried to understand how groupware facilities could take advantage from the recent development of tabletop applications and to contribute to deepen understanding in how social dynamics affect the adoption of collaborative technologies.

In this paper we first present the main features of the Collaborative Workspace. We then describe the user study and, finally, we focus on the two key phenomena emerged from user studies: role specialization and role re-distribution as two aspects contributing to shape individual and group experience around collocated tabletop groupware.

2 The Collaborative Workspace

The design of the Collaborative Workspace was informed by previous users studies and from the guidelines developed in the field of tabletop interaction, in particular [Scott 2003]. Several principles were taken into consideration. First the simplicity, that is, the design of the interface was kept as closer as possible to the affordances offered by traditional artefacts: tables, pen and paper. A paper-based metaphor was used in order to take advantage from the versatility, flexibility and shareability of traditional resources. Second, the CW is meant to encourage adoption by allowing different levels of usage (both at an expert and a at naïve usage of the system). The design followed a principle according to which the system should allow both a low entry level of usage—which requires very few learning—and a higher level, more demanding but providing more sophisticated functionalities. Finally, a balance between single users' requirements and group requirements was pursued. We tried to support users in their individual work as well in collaborative work by providing individual and group spaces as well as individual and group functionalities.

Fig. 1. The Collaborative Workspace

2.1. The system

The Collaborative Workspace functionalities was designed to support four main activities that groups performed during meetings and between subsequent meetings: communication, coordination, storage and connection.

Communication and sharing of information among participants. Roteable virtual documents can be created, oriented, moved and erased by participants by means of the electronic pen. Texts can be edited with a wireless keyboard or with an electronic pen; diagrams and sketches can be drawn using the pen (see Figure 1).

Coordination of teamwork. An electronic agenda was designed to provide support for time scheduling during a meeting. To help improve the temporal organization of the meeting, the agenda items are displayed and can be made

active/inactive. A timer is displayed on the active item and it pauses when the item is made inactive. Using the agenda, pre-prepared documents can be accessed during the meeting. Similarly, documents prepared or modified during the meeting can be associated to specific agenda items in order to be efficiently retrieved after the meeting.
Storage and tracking of activities emerged as key aspects since the beginning of our study and the possibility of saving contents and information produced during meetings proved to be one of the main advantages of the Collaborative Workspace. We distinguish among short-term and long term storage functionalities: a moveable tool called the repository serves to temporarily store and easily retrieve the documentation while the agenda—as explained above—is used as a long-term memory. Furthermore, using the documents uploaded in the agenda and the timing information, the system is able to automatically produce the minutes of the meeting.
Connection between pre-meeting activities with meeting and post-meeting activities. Rather than considering the meeting as an isolated work situation, we situated this event in the wider context of coordination and communication activities taking place before and after the meeting. The system is augmented by a web application that supports users to schedule meetings, inviting other people, uploading documents and pre-setting the agenda; after the meeting, the participants can access the documents produced during the meetings as well as the minutes of the meeting.

3 The user study

After previous formative evaluation studies, we decided to assess the CW by asking small groups of professional to use the system during their naturally occurring meetings. We choose to engage researchers working at our institute to facilitate technical assistance when needed. Given the focus on users' experience, we did not ask users to perform tasks and let them free to accommodate the setting to their specific needs (projectors, whiteboard, laptop).
In this way we reached a sort of "useful approximation" of the system in use in a real context [Reilly, Dearman, Welsman-Dinelle and Inkpen 2005].

3.1. Research questions and methodology

In our previous studies on early prototypes [Zancanaro and Leonardi 2005], we realized that the distribution of activities and roles among participants shape and configure different experiences of usage for the individuals and for the group as a whole. We also noticed an asymmetrical distribution of activities among the participants, with participants using extensively the system during the whole meeting while other participants never interacted with the system. For the present study, the main goal was hence to deepen our understanding on how different working styles–that is, different configurations arising from the distribution of tasks and activities—can affect the user experience. All the meetings were video recorded and several interviews were taken with the participants during the entire study. The analysis of the videos of meetings provided quantitative information on:

- the functionalities used by the participants and amount of usage for each functionality;
- the amount of group usage of the system;

- the amount of individual usage;

Starting from these measures, an observational analysis and interviews with participants were carried out in order to interpret the quantitative data.

3.2. The procedure

The CW was installed in a public meeting room of our institute which was equipped with two cameras and several tabletop microphones.

Three groups—25 people—working in the area of IT accepted to participate in the study. Eight meetings were video-recorded for a total of about 14 hours of system usage (this data does not consider the amount of web site usage). The groups were also observed in a number of meetings before being introduced to the CW.

The groups had different sizes—ranging from three to nine participants—and the meetings were used to carry out different activities. As Table 1 shows, groups had also different level of engagement in the study.

Group	Group size	Meetings goals	Length of the meeting	% system 's usage
Group1	7	Debriefing	2 h	28,5%
	5	Debriefing	40''	69,8%
	7	Debriefing	2 h	17,3%
	9	Debriefing	2 h	19,5%
			Tot. 6 h 40''	
Group2	4	Debriefing and Design	1 h 20''	29,3%
	3	Debriefing and Design	1 h	14,37%
			Tot. 2 h20	
Group3	4	Design	4 h	37,26%
	4	Design	1 h 35''	28%
			Tot. 5 h 35''	

Table 1. Groups, groups size, goals the meetings were planned for, duration of each meeting and percentage of system usage during the meeting.

4. The Collaborative Workspace usage

Observations of meetings carried out before the introduction of the CW underlined some common aspects of the groups' organization and culture. Teams participating in the study were characterized by an informal organization of work and, consequently, of meetings too. No group used special tools to manage teamwork, relying instead on individual artefacts (personal agenda, note-books, laptops) or common office facilities (e-mail, shared folders on intranet) to schedule, circulate documentation and minutes. They also did not use any formal procedures to organize teamwork and relay mainly on informal routines more or less structured. These routines concerned particularly the meeting set up (invitation of participants, choice of the date and choice of the place) and meeting management (keeping up with the agenda during the meeting and preparation of the minutes afterwards). While the three groups employed rather the same procedures to arrange the meetings and to circulate related documentation (mainly by e-mail), the practices of conducting the meetings and of distributing tasks among team members varied significantly and this had important consequences in how they later would use and adapt to the system. For

instance, Group1 used to conduct meetings in a well structured way: each member was in turn appointed by the project manager to summarize the contents of the discussion and to make then circulate the minutes to other participants. The project manager was in charge of the preparation of the agenda and of choosing the date for the meeting. For the other groups, instead, the distribution of the tasks was less structured and emerged during face-to-face interaction.
The nature of the CW requires that the groups learn to use it not in their individual workstations but on a public display. This forced the groups—to different extent—to adjust their work routines. In particular, it imposes a coordination process and a new division of work in order to be able to use the system. As a consequence, the work of some group members rose in complexity while other roles are lightened by the introduction of the system and some tasks decreased in complexity.
We focus in the next sections on different working styles and different modalities of organizing group work around the tabletop groupware.

4.1. Role specialization during meetings with the CW

Rogers and Lindley [2004] argued that tabletop displays encourage group participants toward a more cohesion in working together than the vertical displays and that it foster participants to switch between more roles during group activities. Our early results show that, people tend to stick to specific roles and that the distribution of tasks is quite asymmetrical among group members. This is particularly evident during the first encounters between groups and the CW and in debriefing meetings.
We refer to role specialization to underline a specific way to coordinate group work around the interface. In this case activities are explicitly or implicitly coordinated through a division of labor with participants spontaneously assuming, or accepting to perform specific tasks during the whole meeting. In our case, role specialization was often ironically frozen through nicknames: "the secretary", referring to the person taking notes under dictation, "the trash-man", who takes care of arranging, storing and eventually erasing documents by throwing them in the trash-bin, the "technician", the one who solves technical problems, etc. In some cases, role specialization was the result of an evolutionary trial-and-error process: roles were delegated to, or self-attributed by, individuals who had been successful in performing the given task with the system; these individuals became 'the' experts, acting as mediators between the group and the system. By talking the burden of a given functionality, the expert allowed the rest of the group to keep its attention focused on the task at hand. Often, the expert did not limit his/her activity to operating the system, but often volunteered played a pedagogical role by minimizing technical difficulties and proposing his/her interpretation of the correct usage of the relevant functionality. In other cases, role specialization was not the result of explicit assignment but rather a result of the physical proximity of a person to the functionality needed. For instance, almost all groups used to place the trash bin in a corner of the surface to avoid the risk of inadvertently erasing documents. Since the surface was quite large, often the person organizing the content delegated another person sitting near the trash bin to throw away useless documents.
The consequence of role specialization was often that a high percentage of the CW usage was concentrated on few users. For instance, the data from the Group1 shows,

that 70-80% of the total usage of the system was concentrated on two users. Looking more deeply at the process of task allocation occurring during the meetings itself, we can better understand how activities have been distributed. One of the major concerns of the group was since the beginning to efficiently articulate individual contributions to use the system. This concern clearly arose at the beginning of the first meeting when the project manager asked to other members "should we appoint a coordinator of the system?" With this question, the project manager implicitly suggested a framework for the usage of the system and imprinted the future course of actions. During that first meeting with the CW a lot of effort was spent to negotiate and identify the most efficient working style. Eventually one person was held responsible for writing the minutes while another person managed documents and helped the others participants to suitably use the agenda.
It is worth note that, role specialization was more frequent during the early usages of the CW, and especially in debriefing meetings where the task assigned to, or volunteered for, was that of resuming/taking notes of, the topic discussed.

4.2. Role re-distribution among meetings

If we consider the first meeting of each group, people tend to stick to roles in CW usage for the whole length of the meeting. This is a form "media stickiness" already recognized in other studies [Huysman, Steinfield, Jang, David, Veld, Poot and Mulder 2003]. Yet, if we consider the temporal evolution of usage by considering the individual amount of usage of the CW on different meetings, we notice that participants changed their roles and the usage of the system from a meeting to another. In some case this was due to the perceived workload caused by a simultaneous use of the system while actively participating to the discussion. The problem of balancing attentive resources is well exemplified by Group1. In the first meeting M. used the system–mainly to write summaries and to manage documentation—for 66.6% of the total usage. After the meeting all the participants, speaking about their experience, mentioned the difficulty of balancing group and individual benefits but at the same time they agreed that "it is not a good solution to let the most 'ready and willing' to take always the initiative". Indeed, in the second meeting, the group used another strategy: rather than having one person listening, interpreting and writing the notes for the whole group, they exploited the possibility of sharing a document to collaboratively write the minute once the discussion on each specific topic was concluded. In this way nobody is isolated from the discussion and the final minutes correspond more to the shared understanding of the group instead that being a personal interpretation of the discussion.
Variations in the usage and in role distribution were also due to the progressive identification of the useful features and of the reduction of the less useful ones. Progressively, text documents are less used—or used in a more collaborative way—while the use of the agenda increases over time and become more and more sophisticated with the use of the timing mechanisms and the progressive use of the automatic minutes functionality.

5 Conclusions

The results concerning the stickiness of roles during the same meeting and interchangeability of roles among different meetings seem to confirm the results of other studies in the field of large-display groupware suggesting that this kind of technologies are less amenable to exploration than single-user applications. Even if it is difficult for the participants testing and experiencing the potential uses of the system during a single meeting, people are willing to experiment and explore the functionalities over a longer period of time. In particular, the need to find a balance between individual and group costs and benefits emerged as a key aspect in order to understand the different working styles developed by the three groups. Even if group efficiency could improve with a system like the CW, especially tanks to the possibility to save and easily retrieve digital documents through the web-service and to better manage and control agenda items, benefits at the individual level remain problematic in particular considering the attention needed to keep on with the interaction while using the system and the resources needed to coördinate the CW usage. Some subjects felt frustrated by the load of work imposed by the system and the attention required to coordinate individual contribution toward the system usage. In particular, writing the minutes caused the major difficulties. While the advantages of having a public report are widely recognized from all the members, the practical accomplishment of this task seemed problematic. All the three groups found it difficult to efficiently organize this activity and tried more than one strategy to overcome this problem. Individuals in charge of this task lamented the difficulty in following the meeting discussion, the embarrassment coming from the public nature of the writing "in front of all", the difficulty in selecting the right information and to transcribe it, the fear to give a personal interpretation rather that an "objective" one. Yet, the systems—in particular the functionalities designed to support the writing of the minutes—lead to reduce the steps usually done to write a final report thus increasing the efficiency but at the same time increasing also the complexity of the task. Even if it is risky for an application trying to impose new organizational practices, it would be useful to encourage a cooperative management of this task or to better support the different steps people usually follow to write the minutes of a meeting, that is, the individual writing and elaboration of the notes and, finally, the sharing of the document among team members in order to adjust its content.

References

Huang E. M; Mynatt E. D., Russell D. M., Sue A. E. (2006) Secrets to Success and fatal Flaws: The design of Large-Display Groupware, in *Computer Graphics and Application*, Vol. 26, n. 1, pp. 37-45.

Huysman M., Steinfield C., Jang C.-Y., David K., Veld M. H., Poot J., Mulder I. (2003) Virtual Teams and the Appropriation of Communication Technology: Exploring the Concept of Media Stickiness, in *Computer Supported Cooperative Work,* Vol. 12, N.

Grudin J. (1994) Groupware and Social Dynamics: eight challenges for developers, in *Communication of the ACM,* Vol. 37, Issue 1, pp. 92-105.

O'Hara, K., Perry, M., Churchill, E., and Russel, D. (eds.) (2004) *Public and Situated Displays*. Kluwer Publishers.

Reilly D., Dearman D., Welsman-Dinelle M., Inkpen K. (2005) Evaluating Early Prototypes in Context: Trade-offs, Challenges, and Successes, In *Pervasive Computing*, Vol. 4, Issue 4.

Rogers, Y., Lindley, S. (2004), Collaborating around vertical and horizontal displays: which way is best? In *Interacting With Computers*, 16, 1133-1152.
Rogers, Y., Rodden T. (2004), Configuring Spaces and Surfaces to Support Collaborative Interactions. In O'Hara, K., Perry, M., Churchill, E., and Russel, D. (eds.) *Public and Situated Displays*. Kluwer Publishers (2004), 45-79.
Scott, S.D., Grant, K.D., & Mandryk, R.L. (2003) System Guidelines for Co-located, Collaborative Work on a Tabletop Display. Proceedings of *ECSCW'03*.
Zancanaro M., Leonardi C. (2005) Evaluating Co-Located Technologies through the Lens of Appropriation: A Preliminary Investigation. In Proceedings of the workshop on User-centred Design and Evaluation of Services for Supporting Human-human Communication and Collaboration held in conjunction with *ICMI05*, Trento, October 7, 2005.

Computational Model for Ethnographically Informed Systems Design

Rahat Iqbal[1], Anne James[1], Nazaraf Shah[2], Jacuqes Terken[3]

1 Coventry University, United Kingdom, Department of Computer and Network Systems, Faculty of Computing and Engineering, {r.iqbal, a.james}@coventry.ac.uk
2 University of Essex, Department of Computer Science, Colchester, United Kingdom shahn@essex.ac.uk
3 Eindhoven University of Technology, Department of Industrial Design, 5600 MB Eindhoven, The Netherlands, J.M.B.Terken@tue.nl

Abstract. This paper presents a computational model for ethnographically informed systems design that can support complex and distributed cooperative activities. This model is based on an ethnographic framework consisting of three important dimensions (e.g., distributed coordination, awareness of work and plans and procedure), and the BDI (Belief, Desire and Intention) model of intelligent agents. The ethnographic framework is used to conduct ethnographic analysis and to organise ethnographically driven information into three dimensions, whereas the BDI model allows such information to be mapped upon the underlying concepts of multi-agent systems. The advantage of this model is that it is built upon an adaptation of existing mature and well-understood techniques. By the use of this model, we also address the cognitive aspects of systems design.

1 Introduction

The development of CSCW systems requires a significant understanding of the cooperative work taking place in the world of work. The desire to service this need encapsulates the problems for the traditional forms of requirements capture as CSCW moves beyond the individual user to recognise the socially organised character of work that should be included within the requirements engineering process. To acknowledge the fact that work has a social dimension to it, researchers need to move literally as well as metaphorically from the laboratory to the field and bring sociology in as one of the collection of disciplines which inform both HCI and CSCW. Applying ethnographic methods of investigation can unfold the social aspects of work practices in the real world.

On one hand, ethnographic research yields valuable insights but on the other hand the findings cannot be readily translated into the design of a system. Given the

theoretical precepts of ethnography, it is difficult to move from description to prescription.
The existing corpus of ethnographic research on several complex and large-scale industrial projects has motivated us to investigate the ways in which such effective and rigorous analysis of work organisation can be mapped onto the most suitable computation model that can input ethnographic analysis for the design and development of a system. Learning from our previous experience on ethnographic studies and that of our colleagues, we explore the ways how different dimensions of ethnographic framework (Hughes, O'Brien, Rodden, Rouncefield and Blythin 1997) can directly be mapped onto a BDI model (Rao and Georgeff 1995) of artificial intelligent agent.
The rest of the paper is organised as follows. Section 2 provides an overview of ethnography and intelligent agents. Section 3 discusses our proposed approach which consists of mapping different ethnographic views on to concepts embedded in Multi-agent systems. Section 4 briefly discusses agents and their ability to support humans. Section 5 concludes this paper.

2 Theoretical Foundations

The objective of this research is to propose a computational model by mapping the concepts of ethnographic framework onto a BDI model of artificial intelligent agent. The use of this model will ensure that the user requirements are captured and interpreted according to the most important dimensions of ethnography such as distributed coordination, plans and procedures and awareness of work, and mapped onto the underlying concepts of multi-agent systems (belief, desire and intention) in order to build a system. The advantages of the proposed computation model are that it provides a means of mapping ethnographic concepts onto multi-agent systems. Most importantly both ethnography and multi-agent systems take into account the social aspects of work practices. Multi-agent architectures also provide standard technology and protocols for implementation of these systems.

2.1 Ethnography

The primary focus of ethnography is to unfold the social aspects of work practices and to inform designers about the sociality of work practices taking place in the real world of work. Ethnographic research delivers an extensive amount of data. This provides a rich, textual and concrete exposition of the analysis of work practices in the domain of investigation.
Many researchers and practitioners have found that ethnographic analysis of work settings provide useful insights to the work processes and settings that help system design (Bentley, Hughes, Randall, Rodden, Sawyer, Shapiro and Sommerville 1992; Hughes, King, Rodden and Anderson 1994; Myers, 1999; Shapiro 1994; Crabtree, Hemmings, Rodden and Mariani 2003). Ethnography has successfully been applied to various complex, large scale and domestic workplaces in a number of domains such as domestic and office life, small and large industry, control rooms and financial banks, etc.
The main outcome of these studies is the recognition that computer systems which are developed without any systematic help from the social sciences may not

thoroughly address the needs of the users (Goguen and Linde 1993). Also important is the growing realisation that most systems fail because they do not resonate with the work as it is actually done as a 'real world' and 'real time' phenomenon (Grudin 1994).

2.2 Multi-Agent Systems

An Intelligent Agent is a software entity, which perceives its environment through sensors and performs actions on the environment through effectors. Artificial intelligent agents consist of a number of autonomous interacting software entities called agents, whose behaviour is governed by social norms or interaction protocols. In general a Multi-agent System (MAS) is a computational system in which agents with different capabilities and resources perform their task by coordinating and cooperating with each other in order to achieve a set of goals.
Michael Wooldridge and Nicholas Jennings (Wooldridge and Jennings 1995) argue that the term agent is used to denote a hardware or (more usually) software-based system that should have the following properties:

1 *Autonomy:* Agents operate without the direct intervention of human or others, and have some kind of control over their actions and internal state.
2 *Social Ability:* Agents interact with other agents (and possibly humans) via some kind of agent communication language.
3 *Reactivity:* The agents perceive their environment, and respond in a timely fashion to changes that occur in it.
4 *Pro-activeness:* Agents do not simply act in response to their environment; they are also able to exhibit goal-directed behaviour by taking the initiative.

Agents are built based on an approximation of the human reasoning process encapsulated in BDI architecture. The BDI architecture of an agent is based on the Bratman theory of practical reasoning in humans (Bratman, 1987). It represents three major components of an agent; known as belief, desire and intention. The belief component represents the information the agent has about its environment and its capabilities, desire represents the state of affair the agent want to achieve and intention corresponds to the desires the agent is committed to achieve. BDI architecture is also called deliberative architecture because a BDI based agent is involved in a deliberation process before deciding what actions to take to achieve a particular goal in a given situation (Rao and Georgeff 1995; Rao and Georgeff 1998).
MAS can be viewed as a collection of autonomous problem solving entities, capable of achieving their goals through interaction, coordination, cooperation and collective intelligence. Individual agent possesses limited amount of data processing power and capabilities to solve the problem. Most application domains require multiple cooperative agents for their problem solving.

3 Proposed Approach

We use an effective and rigorous ethnographic framework that helps organise ethnographic research findings. We use three viewpoints for each of ethnography and MAS. In ethnography, these viewpoints are referred to as distributed coordination;

awareness of work; and plans and procedures (Hughes, et al. 1997), whilst in MAS, these viewpoints consist of plans and actions; belief and coordination protocols (Rao and Georgeff 1995).
In the subsequent sections, we discuss different viewpoints of ethnographic model and BDI model of MAS. We map the concepts of ethnographic model on to the BDI model of MAS as shown in Table 1.

Ethnography	Artificial Intelligent Agents	Level of analysis
Distributed coordination	Coordination protocols	Social Behavioural analysis
Plans and procedures	Plans and actions (desire and intension)	Functional analysis
Awareness of work	Belief	Structural analysis

Table 1: Mapping Description

3.1 Distributed Coordination

In ethnography, distributed coordination refers to the fact that the tasks in complex settings are carried out as: part of patterns of activity, operations within the context of division of labour, 'steps' in protracted operations and contributions of continuing 'process' of activity (Hughes, et al. 1997). The activities are dependent upon each other. Distributed coordination involves coordinating the interdependencies between the activities and describes how the tasks are performed.
In the development of MAS, distributed coordination highlights the importance of actions and tasks within the system and describes the manner and means by which work is coordinated. It also emphasises the implications of supporting coordination mechanisms. Coordination in Multi-Agents based CSCW systems is managed by standards coordination protocols (Kuwabara, Ishida and Osato 1995). These protocols are specified in terms of a legal sequence of actions in order to constrain the behaviours of the interacting agents (Yolum and Singh 2002).
As in human society, effective coordination is also essential in MAS in order to achieve commons goals among autonomous agents. The role of coordination is to maintain various forms of interdependencies that occur in a system of interdependent agents. The systems that are capable of solving problems cooperatively must employ standards, or mutually agreed upon ad-hoc coordination mechanisms, in order to manage dependencies among their interrelated activities.
In MAS, coordination deals with the problem of ensuring that a community of individual agents acts in a coherent manner. Nwana and his colleagues emphasise the need of coordination in agent systems for managing, anarchy; global constraints; pooling of distributed expertise and knowledge; and dependencies between agents' actions (Nwana, Lee and Jennings 1996). Several coordination mechanisms have been developed to address the problem of coordination in MAS. These mechanisms range from social laws (Smith 1980) that constrain the acceptable behaviours of agents, to explicit coordination models (Jennings, Sycara and Wooldridge 1998; Durfee, and Montgomery, 1990) and interaction protocols that are used to guide the society's behaviour.

3.2 Plans and Procedures

Within the ethnographic model, plans and procedures provide an important means by which distributed coordination is achieved. A wide range of artefacts such as plans, schedules, manuals of instruction, procedures, job descriptions, formal organisational charts, and workflow diagrams are all examples of plans and procedures which allow people to coordinate their activities (Hughes et al. 1997; Schmidt 1997). Hughes and his colleagues further stated that most important is the understanding of how 'plans and procedures' are used to organise activities. Plans and procedures help to identify the different actors and their functional roles within the working organisation.

In MAS, each agent has its own plan library which can be built using human knowledge. An agent's plan is an implementation of well-defined business functionality or a part of it. Each plan which is stored in a library is a recipe or a set of actions that an agent uses to achieve a particular goal. The plan library provides the flexibility to extend an agent's functionality in a modular way. A BDI agent plan consists of three parts; the plan's context, the plan's relevance and a recipe of action that is executed to achieve a desired goal. A plan is invoked to achieve a goal which is always executed in a given context.

3.3 Awareness of work

The third component of the ethnographic model refers to the way in which the work activities are made available to others (Hughes et al. 1997). The physical layouts of workplaces can affect the ability of people to make reciprocal sense of the others' activities. In the workplace, the visibility or intelligibility takes place through talking aloud as someone works or through the representation of the work to be done. This can be achieved through forms, memos, and worksheets etc. which make obvious the current stage of the work.

Contextual information can be stored in an agent's belief mechanism in order to provide context awareness during execution of its plan. A plan that is valid in one context may not be valid in another context for achieving the same goal.

In BDI agents such as JACK the context aware functionality is implemented by a context method within the plan. A plan's context method is executed in order to determine the applicability of the plan to achieve a given goal in a particular context. Context methods always operate on agents' beliefs to confirm or refute the presence of a given fact.

4 Discussion

There are several ways how agents can support human-human communication and collaboration. It has been proven that agents can support the individual by maintaining a common visual space; support communication among team members by automatically passing information to the relevant team members and support task prioritisation and coordination by maintaining a shared checklist (Sycara and Sukthankar 2006).

In an agent based computation model, information that is needed to perform a user's certain task is implemented. This implementation contains both the data and internal logic which is then used to process the information on behalf of a human. Such implementation contains several alternative plans suitable to perform the task in different situations.

Agents have a varying degree of autonomy depending upon their domain of application. For example, agents may have limited autonomy in applications that involve monetary transactions as compared to information retrieval systems in which agents may have the maximum degree of autonomy. This level of autonomy determines the level of decision making delegated to the agents. This, in fact, shows the division of decision making between agents and humans. In order to reduce cognitive load on humans a certain amount of decision making can be delegated to agents without the need for human intervention. There is a limit to the amount of decision making that can be left to an agent and there may be a situation where human users have to intervene and make their own decision based on information gathered by agents.

When agents receive a task related event, they set a goal to achieve, i.e., the completion of a task. In order to achieve this goal, the agents select the best plan from their plan library to execute, based on their current belief. Based on events, agents initiate their reasoning mechanism in order to select the best available plan. Agents do not simply execute their plans; they apply reasoning to the selection of their plan for a given situation.

Multi-agent systems which are built upon this principle of varying autonomy may reason about the transfer of control decisions and perhaps assume control when the human is not available or unable to do a task. As a matter of fact, the human has greater abilities and task related expertise than the software agents but usually less time. Humans can spend their time on the most important tasks where human intervention is necessary, for example, decision making, while the agents can carry out less important or tedious tasks. Agents can be given a limited amount of decision making attorney.

Decision making in a given task or problem domain is distributed among cooperating agents and humans. Important decisions are delegated to human and less important decisions are delegated to agents. In this way, agents act as assistant of humans and thus help to reduce cognitive load (Ritter and Young 2001) on them by performing tedious activity in a dynamic environment and bringing important events to the attention of humans when a human decision is needed.

5 Conclusions

In this paper, we propose a computation model based on the BDI agent in order to address the problems encapsulated within ethnographic research that limits its usage and adoption by the industry. The model is built upon an adaptation of existing mature and well-understood techniques. The use of this model will ensure that the user requirements are captured and interpreted according to the most important dimensions of ethnography such as distributed coordination, plans and procedures and awareness of work, and mapped onto the underlying concepts of multi-agent systems (belief, desire and intention) in order to build a system. We also believe that the use of multi-agent systems may reduce cognitive load on humans by delegating partial decision making to the agents. This will allow human to reserve their time and efforts for the most important decision where human intervention is important.

References

Hughes, J., O'Brien, J., Rodden, T., Rouncefield, M. and Blythin, S. (1997) Designing with ethnography: A presentation framework for design, in I. McClelland, G. Olson, G. C. van der Veer, A. Henderson and S. Coles (eds.): Designing Interactive Systems: Processes, Practices, Methods and Techniques (Proceedings of DIS'97), Amsterdam, Netherlands, ACM Press, pp.147-58.

Rao A. S. and Georgeff M. P., (1995) BDI Agents: From theory to practice, Proceedings of the First Intl. Conference on Multiagent Systems, San Francisco 1995.

Bentley, R., Hughes, J., Randall, D., Rodden, T., Sawyer, P., Shapiro, D. and Sommerville, I. (1992) Ethnographically informed systems design for Air Traffic Control', in Proceedings of CSCW'92, ed. J. Turner and R. Kraut, 123-129, Oct. 31-Nov 4, Toronto, Canada: ACM Press.

Hughes, J., King, V., Rodden, T. and Anderson, H. (1994) Moving out of the control room: Ethnography in systems design, in R., Furuta and C., Neuwirth (eds.): Computer Supported Cooperative Work (Proceeding of CSCW'94), Chapel Hill, USA, ACM Press, pp..429-39.

Myers, M. D. (1999) Investigating information systems with ethnographic research, Communications of the Association of Information Systems vol. 2 no. 23.

Shapiro D., (1994) The Limits of ethnography: combining social sciences for CSCW, in R., Furuta and C., Neuwirth (eds.): Computer Supported Cooperative Work (Proceeding of CSCW'94), Chapel Hill, USA, ACM Press, pp. 417-428.

Crabtree, A., Hemmings, T., Rodden, T. and Mariani, J., (2003) Informing the development of calendar system for domestic use, in K. Kuutti, E. Karsten, G. Fitzpatrick, P. Dourish and K. Schmidt (eds.): European Computer Supported Cooperative Work (Proceedings of ECSCW'03), Helsinki, Finland, Kluwer Academic Publisher:, pp.119-38.

Goguen, J.A., and Linde, C. (1993) Techniques for Requirements Elicitation, Requirements Engineering, IEEE Computer, pp. 152-164.

Grudin, J., (1994) Eight challenges for developers, Communication of the ACM, vol. 37, no.1, pp. 93-105.

Wooldridge M., and Jennings N. R. (1995) Intelligent agent: theory and practice, in Knowledge Engineering Review, 10(2), pp. 115-152.

Bratman M., (1987) Intentions Plans and Practical Reason. Harvard University Press, Cambridge. 1987.

Rao A. S., and Georgeff M. P., (1998) Decision procedures for BDI logics, Journal of Logic and Computation, 8(3), pp. 293-344.

Kuwabara K., Ishida T., and Osato N., (1995) AgentTalk: Coordination Protocol Description for Multiagent Systems, Proceedings of the First International Conference on Multi-Agent Systems (ICMAS-95), pp. 455, 1995.

Yolum P., Singh M. P., (2002) Commitment machine, Proceedings of the Intelligent Agents VIII, Lecture Notes in Artificial Intelligence, Vol. 2333, Springer Verlag, 2002, pp. 235-247.

Nwana H. S., Lee L., Jennings N. R., (1996) Co-ordination in software agent systems, The British Telecom Technical Journal, 14 (4), 1996, pp.79-88.

Smith, R.G., (1980): 'The contract net protocol', IEEE Transactions on Computers, 1980, pp. 1104-1113.

Jennings N. R., Sycara K. and Wooldridge M., (1998) A roadmap to agent research and Development, autonomous agents and multi-agent systems, 1998, pp. 7-38.

Durfee, E. H., and Montgomery, T. A., (1990) A hierarchical protocol for coordinating multi-agent behaviour, Proceedings of the National Conference on Artificial Intelligence, The AAAI Press, 1990, pp. 86-93.

Schmidt, K., (1997): Of maps and scripts: the status of formal constructs in cooperative work, in S., Hayne, W, Prinz, M., Pendergast, K., Schmidt (eds.): Supporting Group Work: The Integration Challenge (Proceedings of GROUP'97), Arizona, USA, ACM Press, pp. 138-147.

Sycara K, Sukthankar, G., (2006) Literature review of teamwork models, Technical Report, CMU-RI-TR-06050.

Ritter, F., and Young, R., (2001) Embodied models as simulated users: introduction to this special issue on using cognitive models to improve interface design. In Internal Journal of Human Computer Studies (2001) 55, pp. 1-14.

Developing Ubiquitous Collaborating Services in Grid Environment

B Praveen Viswanath[1] *and Kashif Iqbal*[2]

1 University of Bolton, Department of Business Logistics and Information Systems, United Kingdom, bpv1@bolton.ac.uk
2 Universiti Sains Malaysia, School of Computer Science, 11900 Bayan Lepas, Penang, Malaysia, kashif@cs.usm.my

Abstract. Grid computing provides a world largest ubiquitous collaborative environment that allows a number of applications to communicate with each other in order to share computational resources and services seamlessly within this environment. In this paper, we propose a viable execution environment to enable these ubiquitous services, also known as grid services to collaborate through WS-BPEL (Web Services Business Process Execution Language) engine. Grid services are stateful software components running on grids and WS-BPEL is business workflow description language that is useful for coordinating grid services as it is for stateless web services. But WS-BPEL engine is mainly used for running business process workflows by facilitating collaboration of stateless web services, with little or no direct support for grid services collaboration. So, the main focus of this paper is to propose an execution environment that will enable grid services collaboration using WS-BPEL engine, by providing a direct approach that attempts to minimise developer efforts. Finally, a comprehensive grid demonstration application is adopted to illustrate the application of the proposed approach within the execution environment.

1 Introduction

The emergence of new technology to support communication and collaboration between humans, between human and machine and between machines seems the realisation of Mark Weiser's vision (Weiser 1993) of ubiquitous computing. On one hand, small, portable and lightweight devices such as cell phone, PDA (Personal Digital Assistant) and Tablet PC are helping us dramatically to communicate and share resources. On the other hand, grids can be deemed as a large scale ubiquitous computing environment that facilitates a great deal of effective communication and interoperability between various applications and devices.

Grid computing, most simply stated, is distributed computing taken to the next evolutionary level (Berstis 2002). Grids provide an illusion of a fairly simple but at the same time a large and powerful computer that is self managing. This large computer is in fact huge a collection of systems that share a wide range of resources amongst them.

Grid computing provides a platform for developing scientific applications, whereas WS-BPEL (WS-BPEL 2007) is a language used to specify business process workflows. WS-BPEL is a well established industry standard for business process workflows, backed by major industry leaders. WS-BPEL enables users to describe business process activities as Web services and define how they can be connected to accomplish specific tasks (OASIS 2007). Both Grid computing and WS-BPEL are working in isolation. Many researchers have attempted to use WS-BPEL to define business process workflows within the grid environment. Their solutions are not as straightforward with their applications and therefore cannot easily be disguised for usage in grids.

This paper intends to discuss how grid application components use web services technology and most importantly how such applications can benefit from a business process composition language like WS-BPEL, which is primarily used to coordinate business web services. The rest of the paper is organised as follows. Section 2 provides an overview of Grid computing. Section 3 discussed WS-BPEL in brief. Section 4 presents our proposed solution that provides an execution environment for grid computing. Section 5 illustrates the application of the proposed approach. Section 6 concludes this paper.

2 Grid Computing

Grids provide a world largest collaborative environment that allows homogeneous as well as heterogeneous applications to communicate with each other in order to share resources. Some of the properties of grids are as follows: Grids are very large systems consisting of diverse resources spread over a large geographic area. Grids are inherently distributed and involve operations that move data between resources that are considerably critical and substantial. Grids are very dynamic environments wherein the resources available change rapidly and in most cases on the same time scale as the lifespan of a typical application (Yaacob and Iqbal 2003).

Grid computing infrastructure revolves around sharing network, computers and resources associated with them. Grid networks support services that are distributed over large computer networks and these services need to share, coordinate and synchronise (Yaacob and Iqbal 2003) within the scope of the desired grid application. Sharing within grids is centred on being able to provide authorised access to available resources. Resources within grids can be classified as hardware components like memory, hard disk space, bandwidth, application programmes and data. Resources within grids are very dynamic and they need to be maintained and made available with appropriate authentication and authorisation mechanisms.

Grids involve extensive coordination and synchronisation between various activities. A grid application usually divides a programme into independently executable sub-programmes and executes them on multiple machines across the grid network. The whole purpose is to overcome the inherent disability that prevents a single machine from executing a large programme. These sub-programmes need to be well coordinated and synchronised to achieve the application's end objective. Also, grids require execution of services both synchronously and asynchronously. Synchronous execution involves making requests to services and getting immediate responses. Predominantly grid applications are involved in long running asynchronous operations where the application needs to wait for services to complete a long running task before it can move on to the next stage. This type of asynchronous grid application is illustrated in section 5.

3 Web Services Business Process Execution Language

WS-BPEL enables the orchestration of web services by composing a set of web service invocations into a workflow that is executable. A WS-BPEL engine takes responsibility of executing a workflow specified using WS-BPEL and also exposes it as a web service. WS-BPEL was originally devised by combining IBM's Web Service Flow Language and Microsoft XLANG languages. The process of standardising WS-BPEL is currently under the hands of the OASIS WS-BPEL technical committee.

WS-BPEL defines a model and a grammar for describing the behavior of a business process based on interactions between the process and its partners (WS-BPEL 2007). The partners constituting the business process can be interacted with web service interfaces and a partnerLink (WS-BPEL 2007) summarises the relationship structure at the interface level. WS-BPEL coordinates various services to achieve a common business goal. It also incorporates all the necessary logic to achieve this desired goal. WS-BPEL has built-in mechanisms to deal with business exceptions and processing faults (WS-BPEL 2007). It also provides compensation mechanisms in the event of exceptions or reversal of requests by partner services.

WS-BPEL is a combination of various specifications that include WSDL (Web Services Description Language) (WSDL 2001) and XML Schema (XML Schema 2001) for the date model and XPath (XML Path Language) (XPath 2007) and XSLT (Extensible Stylesheet Language Transformation) (XSLT 1999) that support data manipulation. Partner services and other external resources are described by WSDL. A WS-BPEL process can be deployed in different ways and in different scenarios, while maintaining a uniform application-level behavior across all of them (WS-BPEL 2007).

4 Proposed Approach

The proposed approach attempts to illustrate a mechanism via an execution environment to facilitate the collaboration of grid services using WS-BPEL. Globus Toolkit 4 (Globus Toolkit 4.0 2007) is an open source software toolkit used for building grid systems and applications. Globus Toolkit 4 Java WS Core (GT 4.0: Java WS Core 2007) provides an implementation for running stateful web services on the grid, by combining specifications that include WSA (Web Services Addressing) (WSA 2004), WSRF (Web Services Resource Framework) (WSRF 2004) and WSN (Web Services Notification) (WSN 2004). These stateful web services also known as grid services are described via WSDL and can be accessed via standard network protocols like SOAP (Simple Object Access Protocol) (SOAP 2003) over HTTP.

The motivation for coordinating grid services with WS-BPEL was derived from the fact that WS-BPEL supports simple message exchanges between services to complex interaction between services by supporting synchronous and asynchronous, long running, stateful, machine-machine processes (Zager 2005). WS-BPEL provides useful activities like <flow> (WS-BPEL 2007) that facilitates execution of various services in parallel. All these WS-BPEL features fit well with some of the

requirements of grid applications discussed in section 2. Also, the use of WS-BPEL built in mechanism for exceptions and faults mentioned in section 3 is particularly mandatory when dealing with long running asynchronous processes that could terminated abruptly within grid environments.
However, there are issues with integrating WS-BPEL technology with grid services and those are well discussed in (Slominski 2006). This gives rise for a need to have a viable execution environment to enable grid services collaboration through WS-BPEL engines. In this section, we propose an approach that showcases such an execution environment that illustrates the potential of combining Globus Toolkit 4 and WS-BPEL technologies.
A few approaches for enabling grid services collaboration using WS-BPEL have been previously proposed. (Chao, Younas, Griffiths, Awan, Anane and Tsai 2005) details an architecture that wraps grid service clients as web services called proxy web services that can be invoked by a WS-BPEL engine. This approach was based on the then available Globus Toolkit 3. (Zager 2005) uses a different approach that involves manipulation of message headers in order to facilitate communication between grid services and their resources. This approach is based on Globus Toolkit 4.
Our approach here has a more straightforward means that relies on restructuring existing grid services clients and exposing them as services that can be invoked by WS-BPEL engine, by bypassing the conventional SOAP over HTTP means of communicating with a service. In order for a WS-BPEL engine to call a grid service based on this approach, a WSDL document needs to be created that provides an interface for achieving the required functionality of the grid service's Java client. A Java class is then written to provide the required functionality of that grid service's Java client, by adhering to the interface defined in the above mentioned WSDL. Finally, the WSDL and the newly created Java class will form the basis for calling the grid service. Section 4.1 illustrates this approach using a popular WS-BPEL engine, Oracle BPEL Process Manager.

4.1 WS-Invocation Framework and Oracle BPEL Process Manager

The approach here is to use a Java class that provides the required functionality of the grid services client and expose it as a service that can participate within the Oracle BPEL Process Manager WS-BPEL workflow. In order to achieve this there is a need to modify only the service binding (WSDL) (Juric 2005) of that service.
WS-Invocation Framework (WSIF 2001) is the underlying technology that will be used to make this approach work by extending the web services model (Juric 2005). It facilitates the description of a service in WSDL, even though the service being used here (a Java class) does not communicate through SOAP. Thus, WS-Invocation Framework permits the mapping of such services to the actual implementation and protocol (Juric 2005).
So, the abstract description of any partner web service, part of a WS-BPEL workflow can be bound to a corresponding resource that communicates using one of the supported WS-Invocation Framework binding. Since a Java class provides the required functionality of the grid services client, the Java binding of the WS-Invocation Framework facilitated by the Oracle BPEL Process Manager is used. The realisation of this approach is shown in Fig. 1.

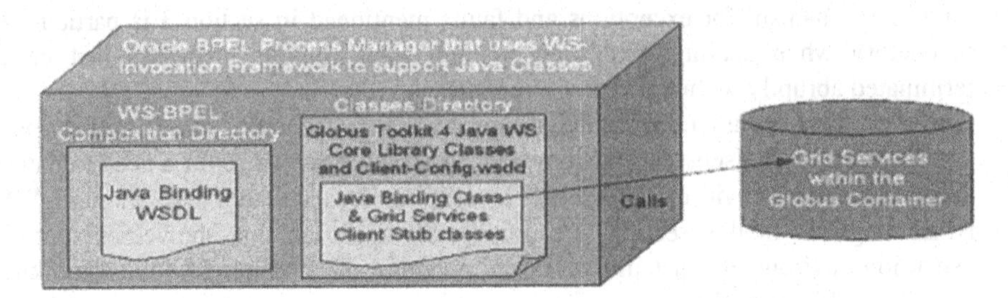

Fig 1. Execution Environment with Oracle BPEL Process Manager

Figure 1 illustrates an execution environment that involves tailoring the Oracle BPEL Process Manager to enable it to call grid services. This can be achieved by:

- Including runtime requirements for grid services client i.e. Java WS Core library classes and client-config.wsdd (client-config.wsdd 2007) within a location on the Oracle BPEL Process Manager's classpath (classes directory).
- Creating the WSDL with Java Bindings for the Web Service to be invoked by the WS-BPEL engine, by providing an interface for achieving the required functionality based on the grid services Java client.
- Creating the Java Binding class, written to provide the required functionality of the grid services Java client by adhering to the interface provided by the above WSDL.
- Copying the Java Binding class and the grid services client stub classes to a location on the Oracle BPEL Process Manager's classpath (classes directory).

4.2 Benefits of using the proposed approach

Existing grid services clients already contain the necessary logic to invoke the required grid services. Thus the approach requires only restructuring of grid services clients, which involves adhering to interfaces that are created to provide the desired functionalities of the grid services clients. The restructuring effort here is very minimal. Also, restructured grid services client programmes still preserve their original intended functionalities. Invoking services through WSIF maintains the performance of native protocols (Juric 2005). Thus, when invoking a native Java class that provides the required functionality of the grid services Java client, the need to pay the performance penalty of web services is eradicated. Since the WS-BPEL composition created on the basis of collaborating grid services is a web service in itself, it can be used for possible hybrid compositions, depending on the nature and requirements of the same.

5 Applying the Proposed Approach

In order to apply the proposed approach, a Grid Demonstration Application (Introduction to Grid Computing 2005) that was built to illustrate some of the functionalities provided by the Globus Toolkit 4 was used. This application is a system that takes Scalable Vector Graphics (SVG) files and uses nodes on a grid to render a set of JPEG files representing sub-images of the complete image (Introduction to Grid Computing 2005).

The demo application consists of three programme components:

1. The RenderClient: This is a Java application with appropriate graphical user interface to drive the rendering work on the grid and display all the rendered sub-images as one composite image.
2. The RenderWorker: This is a java application that converts a sub-image of the SVG file into a JPEG file. Depending on the number of sub-images that would be sent to each node, there could be one or more RenderWorkers running on each node.
3. The RenderSourceService: This is a grid service (just one running on the grid) that is invoked through the RenderClient to give work instructions to the RenderWorker processes on the grid.

The above described RenderClient Java class was stripped off its user interface and restructured to call two grid services:

1. The GRAMLocator grid service that queries the virtual organisation root node on the grid to obtain a list of nodes that are registered and running in the virtual organisation. The GRAMLocator grid service was not part of the Grid Demo Application and was added because this service provides a fundamental functionality that can be used by a wide variety of grid applications.
2. The RenderSourceService described above.

The restructuring of the RenderClient involved splitting it into a client that calls the GRAMLocator grid service and another that uses the RenderSourceService to enable staging of necessary files to each node on the grid so they can run the RenderWorker application that performs sub-image conversions, and retrieve the resultant sub-images from all the participating nodes. These activities within this client are executed using threads as there are a number of participating nodes and each node will have a number of sub-images to convert. Both clients are exposed as services called GRAMLocator and DespatchExecuteRetrieve respectively and is called within a WS-BPEL workflow using the approach detailed in section 4.

The WSDL interface for the WS-BPEL workflow describes all the necessary inputs as described in (Introduction to Grid Computing 2005). In the WS-BPEL workflow, the GramLocator service's method takes the virtual organisation root node as input and returns an array of nodes registered and running in the virtual organisation. The DespatchExecuteRetrieve service's method accepts this array and returns the final job state for each sub-image, all wrapped inside an array. Since this WS-BPEL workflow is a web service in it itself, it can be used for possible hybrid compositions that require the use of a service that accepts SVG files to render a set of JPEG files representing sub-images of the complete image. A service that could be part of the above mentioned hybrid composition is a service that initiated this hybrid composition and wants to retrieve all the rendered JPEG files. The WS-BPEL based grid services collaboration is illustrated in Fig. 2.

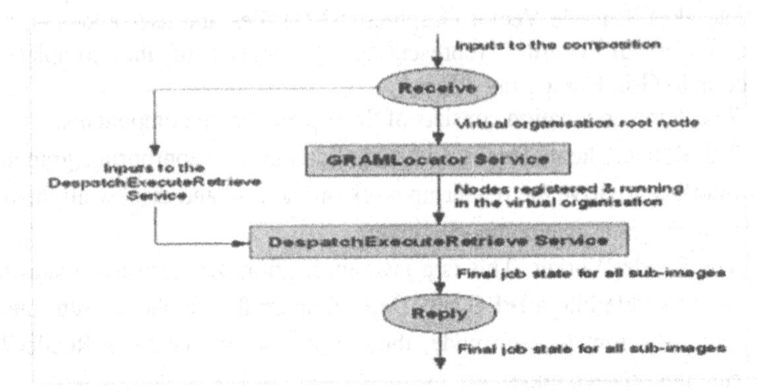

Fig 2. WS-BPEL based Grid Services Collaboration

6 Conclusion

This paper proposes an execution environment that manages ubiquitous collaborating services in grid environment using WS-BPEL. The execution environment was backed by a simple technique of tailoring already existing grid services clients to run as services within WS-BPEL engines. The goal was to minimise changes to existing application software that uses grid services and at the same time benefit from a well established workflow language like WS-BPEL. The execution environments illustrated were specific to Java-based grid services implementations but can be adopted to support other languages that are used to implement grid services like C, Python, and Microsoft languages with WSRF.NET (WSRF.NET 2007). This would require WS-BPEL engines to provide extensive support for invocation handlers that could help expose programmes written in these languages as WS-BPEL invokable services.

References

Weiser, M. (1993) Some Computer Science Issues in Ubiquitous Computing, Communications of the ACM, 36, 75-84.

Berstis, V., (2002) Fundamentals of Grid Computing. Available via: http://www.redbooks.ibm.com/

Chao, K., M.,., Younas, M., Griffiths, N., Awan, I., Anane, R., Tsai, C.F., (2005) Analysis of grid service composition with BPEL4WS, Advanced Information Networking and Applications, AINA.

client-config.wsdd, http://www.globus.org/toolkit/docs/4.0/common/javawscore, April 2007.

Globus Toolkit 4.0 Release Manuals, http://www.globus.org/toolkit/docs/4.0/, April 2007.

GT 4.0: Java WS Core, http://www.globus.org/toolkit/docs/4.0/common/, April 2007.

Introduction to Grid Computing, http://www.redbooks.ibm.com/redbooks/pdfs/sg246778.pdf, December 2005.

Juric, Matjaz B., Using WSIF for Integration, http://www.oracle.com/technology/pub/articles/bpel_cookbook/juric.html, October 2005.

OASIS Web Services Business Process Execution Language (WSBPEL) TC, http://www.oasis-open.org/committees/tc_home.php?wg_abbrev=wsbpel, April 2007.

Slominski A., "On Using BPEL Extensibility to Implement OGSI and WSRF Grid Workflows," Concurrency and Computation: Practice and Experience, 2006.

SOAP Version 1.2, http://www.w3.org/TR/soap/, W3C Recommendation 24 June, 2003.
WSA (Web Services Addressing), http://www.w3.org/Submission/ws-addressing/, W3C Member Submission 10 August, 2004.
WS-BPEL (Web Services Business Process Execution Language) Version 2.0 Committee Specification, http://docs.oasis-open.org/wsbpel/2.0/CS01/wsbpel-v2.0-CS01.pdf, 31 January, 2007.
WSDL (Web Services Description Language) 1.1, http://www.w3.org/TR/wsdl, W3C Note 15 March, 2001.
WSIF (Web Services Invocation Framework), ttp://www.research.ibm.com/people/b/bth/OOWS2001/duftler.pdf, August 9, 2001.
WSN (Publish-Subscribe Notification for Web services) Version 1.0, http://www-128.ibm.com/developerworks/library/ws-pubsub/WS-PubSub.pdf, May 2004.
WSRF (WS-Resource Framework) Version 1.0, http://www-128.ibm.com/developerworks/library/ws-resource/ws-wsrf.pdf, May 2004.
WSRF.NET, http://www.cs.virginia.edu/~gsw2c/wsrf.net.html, April 2007.
XML Schema Part 0: Primer Second Edition, http://www.w3.org/TR/xmlschema-0/, W3C Recommendation 28 October, 2004.
XPath (XML Path Language) 2.0, http://www.w3.org/TR/xpath20/, W3C Recommendation 23 January, 2007.
XSLT (XSL Transformations) Version 1.0, http://www.w3.org/TR/xslt, W3C Recommendation 16 November, 1999.
Yaacob N., and Iqbal R., (2003) Distributed Resource Sharing Technique in Grid Environment, Proceedings of the 9th Asia-Pacific Communication.
Zager, M., (2005) SOA/Web Services - Business Process Orchestration with BPEL: BPEL supports time-critical decision making. http://webservices.sys-con.com/read/

Structuring Community Care using Multi-Agent Systems

Martin D Beer

Communications and Computing Research Centre, Sheffield Hallam University, m.beer@shu.ac.uk

Abstract. Community care is a complex operation that requires the interaction of large numbers of dedicated individuals, managed by an equally wide range of organisations. They are also by their nature highly mobile and flexible, moving between clients in whatever order is necessary to provide both scheduled and unscheduled services. What is important to the person receiving care is that they receive what they expect regularly, reliably and when they expect to receive it. Current systems are heavily provider focused on providing the scheduled care with as high apparent cost effectiveness as possible. Unfortunately, the lack of focus on the client often leads to inflexibility with expensive services being provided when they are not needed, large scale duplication of effort or inadequate flexibility to change the care regime to meet changing circumstances. Add to this the problems associated with the lack of integration of emergency and routing care and the extensive support given by friends and family and many opportunities exist to improve both the levels of support and the efficiency of care.
The move towards Individual Care Plans requires much closer monitoring to ensure that the care specified for each individual is actually delivered and when linked with smart home technology in conjunction with appropriate sensors allows a much richer range of services to be offered which can be customised to meet the needs of each individual, giving them the assurance to continue to live independently.

1 Introduction

Community Care is provided by a complex set of interacting organisations and services (McDonald 1999) that provide a useful platform on which to study the theoretical and practical requirements of a highly mobile workforce that needs to act both proactively and reactively to the needs of the client. This is because of a shift in focus in the services provided to older and chronically sick people from institutional care to care in the community. Community Care is typically provided by a range of independent organizations and agencies, each needing to meet its own targets and objectives within a much broader service framework (Fisk 1989). This often leads to serious service inefficiencies in what is of necessity a highly distributed, fast changing and mobile environment as there are inadequate systems in place to share relevant information without compromising the security of the information held. Another factor is that a considerable amount of community support is provided by informal carers who are generally excluded from the general care management system because of difficulties in integrating them without breaching the official confidentiality requirements.

Agent technology, and in particular mobile agents, provides a means by which effective co-operation (information sharing and communication between autonomous information systems) can take place without compromising the security of the client and the agencies involved, particularly in the highly volatile environment of community care.

Since each agent has complete autonomy it can respond according to the rules of the organization it represents, providing an effective and assured guardian that is totally under that organization's control. No other architectural framework for Distributed Information Systems gives this capability without serious reliability problems.

2 The Scenario

The motivation for developing the scenario was initially to test a model which had been developed to characterise community care throughout Europe (Lunn, Sixsmith,Lindsay, Vaarama, 2003). This model was intended to bring together the different delivery models used in various European counties and offered an opportunity to investigate the capacity of agent-based design to provide an effective means of delivering an enhanced level of service considerably more efficiently and at a much lower cost. This paper discusses the use of semantic agents to service the use cases that handle routine (See Figure 2) and Emergency (See Figure 3}) as described in (Beer, Bench and Sixsmith 1999b) over realistically configured networks. These use cases form a key part of the provision of community care and are often treated separately, leading to massive inefficiencies and duplication. For our purposes in this scenario, we define routine care as the provision of a specified care package on a regular or routine basis based on the provisions of the Individual Care Plan or other recognized agreements and Emergency Care as the unscheduled response alarms and events, however triggered.

The setting for the scenario is that of the moderately sized town of Axebridge, with its own Social Services Department and a number of suppliers of different types of care and medical services. The Emergency services each have single control centres that cover the town. The system therefore needs to service several thousand home units, several care providers who deploy several hundred professional carers each with specific capabilities, a number of general practitioner surgeries, each of which provide medical and nursing services to their own patients and a number of pharmacies who dispense medicines prescribed by any of the doctors practicing in the town. This information is used to design the basic agents and to define the conversation classes that control the interactions between them.

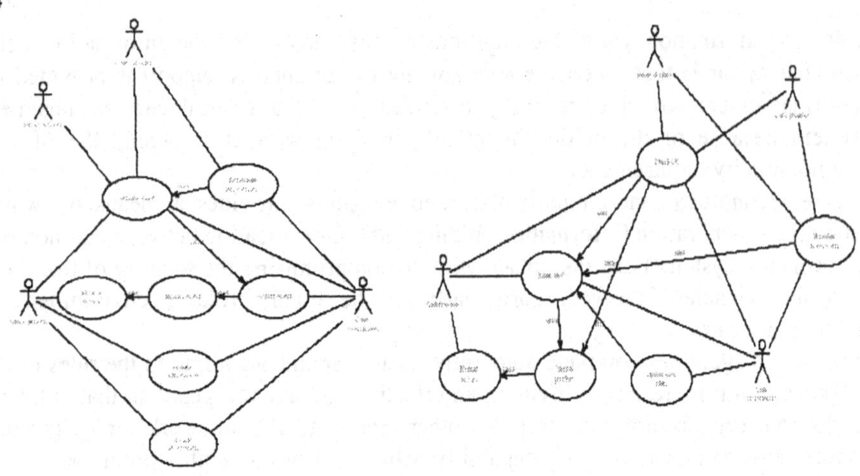

Fig 1. The Routine Care Use Case **Fig 2.** The Emergency Care Use Case

3 The Architecture of the Demonstrator

The architecture is structured so as to filter messages as they pass through various layers so that only relevant information is forwarded on, reducing the communication overhead and meeting the needs for privacy unless it is essential that the information is shared. Care providers are also able to share information, reducing the likelihood of duplication of services, particularly in emergency care situations, where traditionally coordination has not been very strong. Even a minimal level of co-ordination can therefore be extremely valuable.

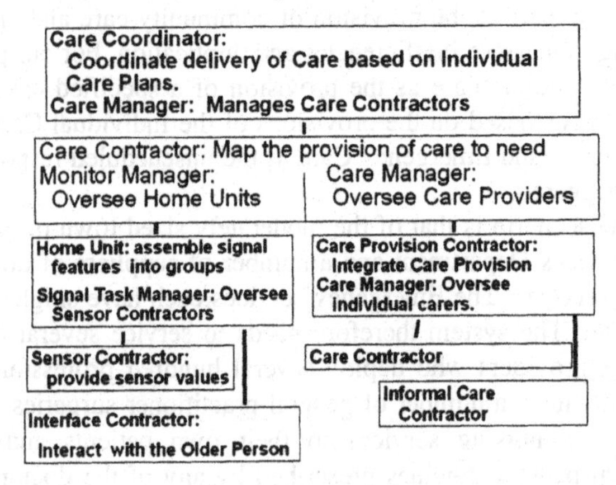

Fig 3. Infrastructure

Figure 3shows the structural relationships between the different types of agent specified within the scenario. These include:

1. the Home Unit and its associated sensor agents
2. the care coordinators that develop and monitor the Individual Care Plans, allocate resources and manage the transfer of financial resources on the basis of the supply contracts etc.

3. the care providers who manage these resources and employ the individual carers, and
4. the carers who actually provide the care.

The interactions between these agents are represented in Figure 3.
In addition, individual carers can communicate with the home units which they are due to visit directly to provide updates on expected arrival times and to obtain additional monitoring information, for example to ensure that the situation is stable until their expected arrival. An example of this would be where a client has fallen and is unable to get up. The client is otherwise unharmed and comfortable and a carer is due later in the day. That carer rearranges their schedule to arrive as soon as practical and their carer agent monitors the client through their Home Unit to ensure that the situation does not deteriorate, by for example the client getting very cold, in the period before help can arrive.
All agents are implemented on the JADE platform (Bellifemine, Poggi and Rimassa 1999) using the Semantic Agent add-on (Louis and Martinez (2005).

3.1 The Home Unit

Each Home unit agent has a number of sensor and alarm agents associated with it. At present these are of two types:

1. Alarm buttons which represent the red buttons provided with alarm systems, and can be both mobile and fixed. All available alarm buttons subscribe immediately to their associated home unit and can only be reset from there once an appropriate response is received.
2. Temperature sensors that are used to ensure that the environment within the home is appropriate. At present, these are associated with particular rooms and are monitored according to a preset schedule so that for example the sitting room is monitored during the day and the bedroom during the night. Alarms of different levels are raised depending on both the level of discrepancy from the ideal temperature and the time that this discrepancy has existed

Further sensors can be added relatively easily as required to test further scenarios.
The Home Unit also maintains a list of carers due to visit (and also whether a carer is currently present) based on semantic information passed to it by the carers as they accept appointments. Should circumstances change, and a particular piece of care is no longer needed, the Home Unit cancels the appointment with the appropriate carer agent, which then notifies its related Care Provider agent that it is now free to take on other commitments.

3.2 Care Coordinators

The Care Coordinator agents monitor the delivery of care against the Individual Care Plans and the contractual arrangements with the various Care Providers. The transactional framework is used to select appropriate care providers when necessary, and to maintain economic balance by ensuring that those services that should be recharged are charged appropriately. As discussed elsewhere (Hill, Polovina, Beer,

2005), this can be a complicated process as often the charging mechanism is not direct in that whether or not the client is charged will depend on factors such as:

1. their ability to pay
2. the service that they receive (in the UK for example, medical services are generally free at the point of delivery but social services are chargeable at standardized rates)
3. the contractual arrangements with the various care providers and care coordinators
4. whether informal or professional carers provide the necessary care

When an alarm is raised the care coordinator agents need to:

1. identify the capabilities required and the speed of response necessary to respond to it based on the specifications given in the appropriate Individual Care Plan
2. negotiate with the various care providers as to who has the capacity to deliver the necessary care within the required time frame
3. select a carer and inform both them and the client of the arrangements
4. monitor the delivery of both emergency and routine care to ensure that it is actually delivered, and in case of non-delivery restart the process with updated requirements
5. manage the transactional arrangements so that the financial provisions of the care delivery are fully complied with

All these are complex tasks that are undertaken by a collection of agents located in a central location that has the necessary communication links with the more localized agents. This is in accordance with the existing organizational models that rely on centralized administrative services to undertake these tasks.

3.3 Carers and Care Providers

It is assumed that carers are available for set one hour slots throughout the day. No account is currently taken of traveling time, which is assumed to be included in the slots. While this is highly ineffective in a practical sense as a five minute visit would be booked for a whole hour, the demonstrator is intended to investigate the effectiveness of the transaction model at various load levels and to investigate the effects of capacity constraints. This means that each carer is available for a maximum of twenty four slots, for each of which the carer agent maintains the following properties:

1. whether the carer is available or not available
2. whether the carer is already booked or not
3. the home (by means of the home agent identifier) that the carer is to visit and the capability required
4. the capabilities of its carer

These give the carer agent sufficient information for the carer agent to maintain an accurate diary for the carer. Pre-arranged visits are loaded from configuration files. These are displayed and can be updated by means of a graphical user interface, as shown in Figure \ref{fig:sally}.

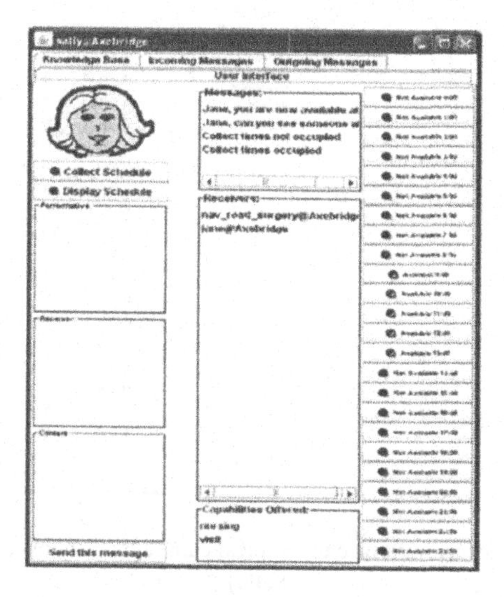

Fig 4. An example of the Home Agent

Carer agents are modeled to be mobile, and follow the activities of the carer to which they are assigned. They subscribe to their appropriate Care Provider so that they can use their semantic capabilities to update the care provider on their diary as it changes. The Care Provider agent can therefore maintain a full list of availability and capabilities that it can provide, and subscribe appropriately (see Figure 5). When a request comes for care, the Care Provider bids if it has that capability available within reasonable time on the following basis:

1. If a carer is already due to visit on the basis of 24/n where n=the number of slots before the carer is due (n=1 means the next slot). This ensures that the first visiting carer with the appropriate capability accepts the commitment
2. If no carer is scheduled then on the basis of p*24/n where p is the nominal cost of that carer for the next available carer with the required capabilities. This allows the care provider to select the cheaper carer with a longer delay, if that is economically efficient
3. If no carer with the appropriate capabilities is available, then on the basis of m*p/n where m is the number of appointments that need to be rearranged + 1 to provide a means of meeting otherwise impossible requests

If the Care Provider's bid is accepted then it negotiates with its associated carer agents to make the necessary bookings and confirms the carer with the relevant Home Unit. Should it not be possible to fulfill the commitment at any stage, it is the care provider agent's responsibility either to arrange for another carer agent to take it on or to notify the appropriate care coordinator agent so that it can be reallocated.

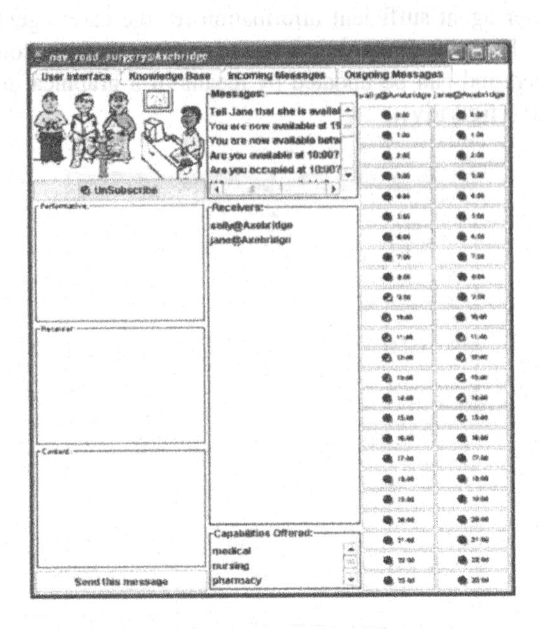

Fig. 5. An Example of the interface to a typical carer agent (in this case the District Nurse, Sally)

The difficulty comes with informal carers who do not fit into this organizational structure. They can communicate directly with the Care Coordinator but this reduces flexibility, where for example a family of relatives who share the responsibility for caring. They can be treated as a small care provider allowing negotiation to be undertaken between family members' agents as to who will respond to the client's needs.

4 Results and Further Work

The original specification of the scenario relied heavily on the FIPA view of agents as independent autonomous computational entities. This has been found to be much too simplistic as the number of individual agents started to multiply very rapidly. It also does not fit well with the underlying care model proposed by (Lunn et al. 2003) For example, each sensor or interface within the home is modeled as a separate agent. If every agent was to allow interaction with every other agent, the necessary discovery services rapidly become unmanageable. The structure represented in Figure 3 provides a means by which agent communities can be segmented into individual zones and individual agent discovery services are then associated with each of them. It is sensible, for example, for each Home Unit to discover and manage the sensors and alarms associated with the home in which it is situated. It is however highly undesirable that it should try to manage sensors associated with another home. This is analogous to each house in a suburban street installing a wireless network and wishing to restrict access to only those computers used by members of that household. This is currently a hot topic (see for example (Omair, Ali, Ahmad, Suguri

2005) and we are currently investigating the most effective ways to better match the underlying agent paradigm to the needs of actual usage scenarios.

References

Beer, M. D., Bench-Capon, T., and Sixsmith, A. (1999) The Delivery of Effective Integrated Community Care with the aid of Agents, Proceedings of ICSC, Lecture Notes in Computer Science 1749, Springer-Verlag pp.303-398

Hill, R., Polovina, S., Beer, M., (2005), Managing Community Health-care Information in a Multi-Agent System Environment, Multi-Agent Systems for Medicine, Computational Biology and Bioinformatics, Utrecht, Netherlands, pp. 35-49.

Bellifemine, F., Poggi, A., and Rimassa, G., (1999) JADE - A FIPA-compliant agent framework. In Proceedings of the 4th International Conference on the Practical Applications of Agents and Multi-Agent Systems, pp 97-108, The Practical Application Company Ltd.

Fisk, M. (1989) Alarm Systems and Elderly People Planning Exchange, London.

Haigh, K. Z., Phelps, J. Geib, C. W., (2002) An Open Agent Architecture for Assisting Elder Independence, in The First International Joint Conference on Autonomous Agents and MultiAgent Systems, pages 578-586.

Mikko Laukkanen, Heikki Helin, Heimo Laamanen (2002) Supporting nomadic agent-based applications in the FIPA agent architecture, Proceedings of the first international joint conference on Autonomous agents and multiagent systems, pp565 - 566.

Louis,V. and Martinez, T (2005) An operational model for the FIPA-ACL semantics, Agent Communication Workshop, Utrecht NL.

Lunn, K., Sixsmith, A., Lindsay, A., Vaarama, M., (2003) Traceability in requirements through process modelling, applied to social care applications, Information & Software Technology, 1045-1052.

McDonald, A. (1999) Understanding Community Care, MacMillan Press Ltd., London, UK.

Omair S., M., Ali, A. Ahmad, F., H., Suguri, H. (2005) AgentWeb Gateway - a middleware for dynamic integration of multi agent system and Web services framework", 14th IEEE International Workshops on Enabling Technologies: Infrastructure for Collaborative Enterprise, pp. 267-268.

Supporting the Social Dynamics of Meetings Using Peer-to-Peer Architecture

Phil Thompson[1] *and Rahat Iqbal*[2]

1 Coventry University, UK, Department of Knowledge and Information Management, p.thompson@coventry.ac.uk
2 Coventry University, UK, Department of Computer and Network Systems, r.iqbal@coventry.ac.uk

Abstract. Most of the time useful output from a meeting is dependent on how comfortable the participants feel in the social environment within which the meeting takes place. This can be enhanced by the person chairing the meeting who on recognising where problems are surfacing within the group can take some kind of action to counter the problem. Studies have been performed using computer aids to assist the person chairing or facilitating the meeting but this does depend on the judgement of one person to be able to make the correct decision and apply the right countermeasures. This paper explores the concept of the participants exercising control over the meeting by using hand-held computers or some other electronic device to signal to the facilitator regarding the social aspects of the meeting. This signal will be picked up by electronic agents which make use of the multiple inputs to directly intervene with the progress of the meeting or bring the matter to the attention of the chair.

1 Introduction

There is a general realization that in the business environment much time is spent in meetings (Ellis and Barthelmess 2003). If those meetings are not managed effectively a lot of time can be wasted. If there are a lot of participants in the meeting and those participants are senior managers of an organization then the amount of money that is being spent achieving very little can be prohibitive. Researchers and business gurus alike have recognized that a good meeting leader, such as a chairperson or facilitator with the right skills can manage the factors that lead to unproductive meetings and enable the objectives of the meeting to be met. The chairperson though often has input into the meeting and is recognized as being a contributor to the topic being discussed. What has become more popular in large organizations is the use of a facilitator (Miranda and Bostrom 1999) who manages the conduct of the meeting but takes no active part in the topics being discussed. A good facilitator can make meetings more productive by keeping the meeting on track, ensuring everybody is able to make a contribution to the meeting, preventing conflict, or suggesting breaks when the interest level of the participants is lacking.

To have a facilitator present at all meetings who is trained in all the skills required can begin to remove the benefits that they can bring. Because the person plays no active part in the meeting they are an overhead and can only be justified if the benefits obtained from using them outweighs the cost. This probably limits their use to meetings between executives in large organisations. To overcome this problem studies have been performed which use computing technology to replace or assist (Bostrom, Anson, and Clawson 1993) the facilitator. Research has been performed (Iqbal, Sturm, Kulyk, Wang, Terken 2005; Sturm, Iqbal, Kulyk, Wang, Terken 2005) which attempts to monitor the speech patterns of the participants in order to assess the progress of the meeting. Technical problems with natural language have so far prevented these from being a real success (Ellis and Barthelmess 2003).

Considering the meeting from the point of view of the participants of the meeting social dynamics becomes important. The members of the group in the meeting need to feel comfortable in order to contribute (Sturm, et al. 2005). There are many factors at work here including the ability to break into an existing dialogue in order to make a point, whether the contributor feels that their point is accepted by the other members or whether participants feel unable to contribute because of being of junior status compared to the others. Studies have been performed where members are able using pre-arranged signals like hand gestures to gain the attention of the facilitator so that the various factors considered above can be accommodated in order to make the meeting more productive.

In order to address these issues this paper proposes an approach which examines the social dynamics of a meeting situation, considers the actions a meeting leader needs to take in order to manage the meeting to a successful conclusion, and uses the participants themselves assisted by computer technology to assist the leader in this task. The rest of this paper is organized as follows. Section 2 examines the social dynamics of a meeting from the point of view of the participants. Section 3 discusses the job of the meeting leader and how the skills of this person enable the meeting to be managed effectively. Section 4 proposes an outline logical design for a prototype which will enable the participants of the meeting to communicate with the leader or even intervene directly in to the proceedings of the meeting if the leader does not respond. Section 5 describes the technical design of the prototype. Section 6 provides conclusions and outlines our future work.

2 Social Dynamics of the Meeting

With respect to social dynamics of the meeting, three basic elements for good interpersonal relations have been proposed (Schutz 1958). These are inclusion, control and affection. Schutz goes further to describe how each of these elements manifest themselves in feelings and fears.

For example inclusion describes behaviour which influences the extent to which individuals feel part of a group and whether they perceive themselves as “in” or “out”. The fear is that they will be excluded. An individual in a group meeting situation who is not allowed to make a point because the other members do not invite

him into the discussion will feel excluded. This according to Schutz will invoke "defence mechanisms" to protect the self esteem of the ignored member. This could result in inviting conflict with other members of the group, deliberately refusing to try to contribute towards the proceedings or could even result in the person leaving the meeting.
Control describes behaviour which challenges the competence of a person and the fear is that the individual will be made to feel humiliated in front of other participants. Considering the group meeting equivalent. If an individual makes a contribution which is badly received by the rest of the group then the contributor will feel humiliated. Especially if shown to be lacking in experience or knowledge, in front of the rest of the group. Again this will result in defence mechanisms coming in to play and the slighted member could withdraw co-operation as before.
The final element, openness describes behaviour which makes the person feel liked by other participants and the corresponding fear is rejection. In a group meeting this could arise as a result of the body language or responses of other members of the group towards one of the member which that member perceives as unfriendly. As before defence mechanisms may then bring about withdrawal from the proceedings by that member.
The work by Schutz described above concentrated on the behaviours and feelings of individuals in respect of inter personal relations and group dynamics but other researchers have focused on the physical conditions which influence the productivity of individuals in a group situation. In a group meeting physical factors during the proceedings such as excessive heat or cold, background noise, onset of tiredness, need for a comfort break will also affect the performance of the members and influence the success of the meeting if they are not rectified.

3 The Rhythm of the Meeting

Recognising and responding to all of the factors described above which can affect the performance of participants of a meeting and the effectiveness of the outcome is the job of the meeting leader, chairperson or facilitator. The person in this role is expected to control both task and relationship in the meeting. The task centred role is to ensure that the purpose of the meeting is achieved and that sufficient discussion takes place with the right technical input to make this possible. The relationship centred role is to make sure that the meeting takes place in a social atmosphere where good group dynamics can be achieved. With good group dynamics there are only problems in the technical discussion of the meeting that will prevent a successful outcome.
To achieve good group dynamics the meeting leader must ensure that relationships between participants at the meeting are kept at a level where good discussion can take place. All of the participants in the meeting should be allowed to make their contribution, no one person should monopolise the discussion. The leader should recognize when the discussion is getting too technical and beyond most of those present. When this happens the leader should ask for a simplified statement. At particular points in the meeting the leader should summarise what has been said so that everybody is following the points being made. There may be points where a rest

is called for and the leader will call for a break while participants recover before continuing. The discussion should not get personal in order to prevent conflict and the leader should de-personalise any disagreement that takes place.
The problem is that it is possible for the meeting leader to be unaware of mounting discontent among the participants in the meeting. The use of computer technology allows the participants at the meeting to keep the leader aware when they feel unable to make a contribution or are in some way unhappy with the conduct of the meeting and their part in it. Also they can do this unknown to the other participants in the meeting who may be influencing their ability to make a contribution.

3 Prototype Design

We designed a prototype that provides feedback to the meeting leader in real time which is directly related to the social dynamics of the meeting. This allows the meeting participants to keep the leader aware of any mounting discontent The prototype system will run on portable devices such as PDAs and laptops.
Each member of the meeting will have one of these PDAs. The proposed interface is shown at figure 1. As the meeting progresses the meeting participants will be able to send signals which will be picked up by the software on the leader's computer.

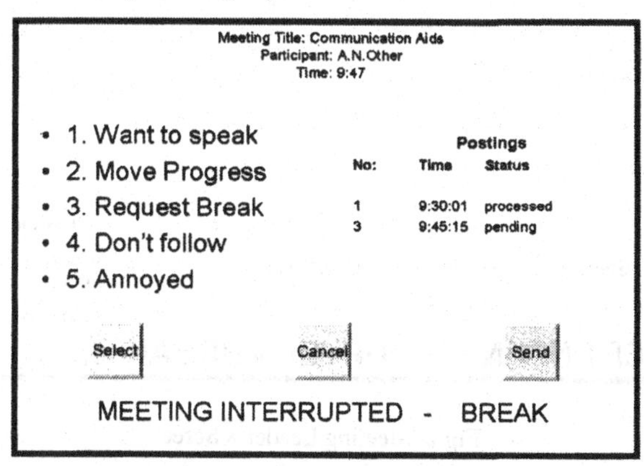

Fig 1. Participants Screen

If the participant feels that an issue has been discussed for too long for example then the "move progress" message can be selected on the PDA and the button pressed to send the message. The message "move progress" will then display on the leader's screen with a number "1" by the side of it indicating that one person has signaled. The leader can choose to respond or not to the message. If another participant sends the same message then the number will increase by one to "2" to indicate that two participants have sent the message. A limit can be set for a message to determine when the leader gets an audible alarm. If the leader continues to ignore the message after a certain preset time all the participants devices will give an audible alarm which is a signal to stop the discussion immediately and for the leader to summarise the discussion and pass on to the next item on the agenda. Similar messages could include: "request a break"; "want to speak"; "don't understand". The PDAs-held devices are capable of storing many of these messages each which couldhave the limits previously described set for them on the leader's computer. Other functionality has not been discussed owing to space limitation.

The leader of the meeting will have a laptop computer with a different display as shown in figure 2. The display on the meeting leaders screen indicates that the message that had been sent, the number of participants who had signaled the message and the initials of the person(s) who had signaled. This enables the leader to respond to the particular participant(s) as necessary. The message is green when first sent, but changes to yellow when the limit of participants was reached and then red when the time limit was passed.

Meeting Title: Communication Aids
Participants: 8
Time: 9:47

Time	Message	From	Status	Count	Limit	Time-Left
9.30.01	want to speak	ANO	ended			
9.40.06	want to speak	ABC	ended			
9.42.06	request break	ANO	pending	4	3	0
9.46.02	want to speak	*	pending	3	4	

Select Reply Delete

Messages

Received	10
Pending	2
Count > 1	1
Count > Limit	0
On Clock	0
Interruptions	1

MEETING INTERRUPTED - BREAK

Fig 2. Meeting Leader's Screen

Each of the PDAs need to be configured with all the relevant messages to be able to manage the meeting effectively to prevent the "defence mechanisms" referred to above coming in to play and possible disruption of the meeting occurring. The name of the participant can be entered into the device which will be displayed alongside the message on the meeting leaders screen. This can be suppressed if required. For example the "want to speak " message would need the name displayed so that the meeting leader could cue the individual concerned but senders of other messages such as "request break" would want to remain anonymous. This would be adjusted on the basis of user studies which would confirm the need for such a facility. The leader's computer can be configured to process the equivalent message types and assign the appropriate alarm and time limits.

4 Peer-to-Peer Architecture

The logical design level previously described will form the application level and this section will describe the transport level (see figure 3). The transport level will be responsible for the transporting of messages to and from the participants' and meeting leader's computers. The processing at this level will be performed by agents using a peer-to-peer architecture. In the peer-to-peer environment agents communicate and collaborate with each other using standard communications languages and protocols.

Each participant will have a participator agent and the meeting leader will have a facilitator agent. The participant agent role will be to process outgoing message requests from the participant and process incoming messages from the facilitator agent. Outgoing messages from each participant will be received by its agent and sent to the facilitator agent. Incoming messages from the facilitator agent will only take the form of "time expired" messages which will initiate the audible alarm on the participant's PDA to interrupt the progress of the meeting.

The facilitator agent will be waiting for incoming messages and on receipt will display the message on the leader's screen, update the message count and if the message count has reached its limit, initiate the message clock process. The message clock process will continue to execute until the time limit for the message has expired and then interrupt the facilitator agent which will send "time expired" messages to each of the participant agents.

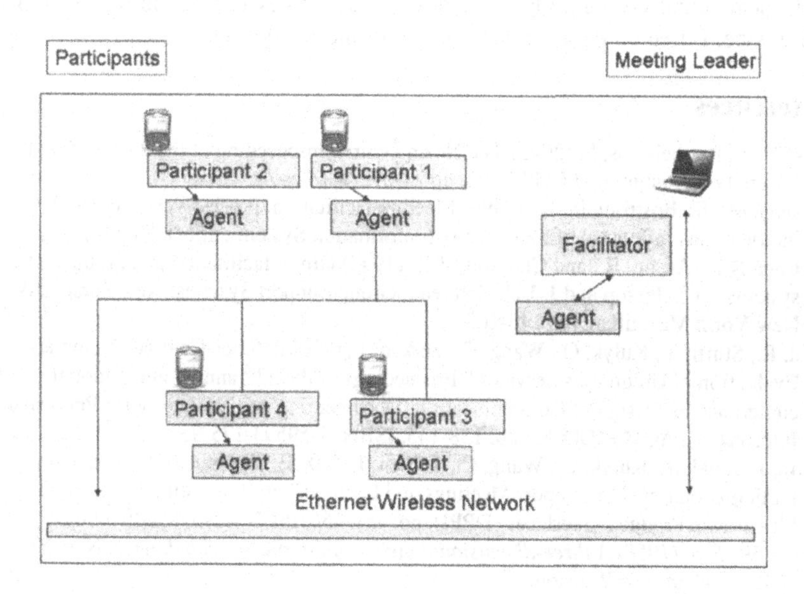

Fig 3. The Meeting Environment

The infrastructure for the transport layer will be a peer-to-peer network. Each participator agent will be a peer on the network which actually resides on the participant's PDA. The facilitator agent will also be a peer which will reside on the meeting leaders's computer. At the physical layer the devices will operate wirelessly using Ethernet protocol, each device being assigned an IP address.

5 Conclusions and Future Work

There is a need to be able to monitor the social environment in meetings otherwise the participants can become disaffected and either withdraw from making contributions, become obstructive or leave the meeting. This can happen as discussed above when the participant is unable to get into the discussion, or because a point is badly received, or if there is not a friendly atmosphere in which points can be made. It is the job of the meeting leader to ensure that participants are kept happy so that they are as productive as possible. If the participants do not make clear that they have a problem because of a reluctance to assert themselves in front of their peers or their superiors then the meeting leader will not be able to do anything about the problem. By giving the participants the means of signalling to the meeting leader unknown to any of the other participants at the meeting this situation can be prevented. The means of signalling is provided by each participant being given a PDA which communicates with the meeting leaders own computer. The meeting leader will then be kept aware of situations which need attention and be able to respond to the signals from the participants at the meeting. If the meeting leader does not respond then the meeting will be interrupted until the problem is resolved.

In the near future we are hoping to conduct a user study to find out the usefulness of this service. Based on the feedback we will refine our design.

References

Ellis C. and Barthelmess, P. (2003) The Neem Dram in proceeding of the 23rd conference on Diversity in computing (TAPIA 03), pp 23-29, ACM press.

Miranda, M., S., Bostrom, P., R., (1999) Meeting facilitation: process versus content interventions in Journal of Management Information Systems, 15(4), 89-114.

Bostrom, R.P.; Anson, R.; and Clawson, V.K. (1993) Group facilitation and group support systems. In L. Jessup and J. Valacich (eds.)Group Support Systems: New Perspectives. New York: Macmillan, pp. 146-168.

Iqbal, R., Sturm, J., Kulyk, O., Wang, C., Terken, J., (2005): "User-Centred Design and Evaluation of Ubiquitous Services", Proceedings of the 23rd annual international conference on Design of Communication: Documenting and Designing for Pervasive Information, ACM SIGDOC, pp. 138-145, ISBN: 1-59593-175-9.

Sturm, J., Iqbal, R., Kulyk, O., Wang, C., Terken, J., (2005): "Peripheral Feedback on Participation Level to Support Meetings and Lectures", in Proceedings of Designing Pleasurable Products Interfaces (DPPI), pp. 451-466, ISBN: 13-9789086580017.

Schutz, W. *Firo (1958) A three-dimensional theory of interpersonal behavior.* New York: Holt, Rinehart, and Winston.

The Chawton House Project: Co-Designing Situated UbiComp

John Halloran[1] *and Eva Hornecker*[2]

1 Department of Creative Computing, Coventry University, Coventry CV1 5FB, UK
John.Halloran@coventry.ac.uk
2 Pervasive Interaction Lab, Open University, Milton Keynes MK7 6AA, UK
eva@ehornecker.de

Abstract. The rise of ubiquitous computing (UbiComp), where pervasive, wireless and disappearing technologies offer hitherto unavailable means of supporting activity, opens up new issues and challenges for co-design. These include the novelty and complexity of UbiComp, the need to consider not just computing but also its environment, and how to promote appropriate user understandings of the technology during the design process. Here we reflect on a case study in designing UbiComp. We discuss a range of co-design activities carried out with the curators of a historic English house, as well as a nearby primary school, to create an educational fieldtrip for children.

1 Introduction

Weiser's vision for the future of computing was of 'ubiquitous' computing ('UbiComp') that 'moves beyond the desktop' to integrate seamlessly with the environment, augmenting a range of human activities be these work, play or learning, as well as everyday tasks like wayfinding and travelling. There are many examples of UbiComp which show how computing responds to and transforms environments. These include Weiser's own work, which aimed at transforming work through the use of new kinds of device interaction ('tabs', 'pads' and 'boards') to allow digital work to be associated with people and their movements and needs (particularly collaborative needs) rather than localised machines (Weiser 1991). Other projects show how UbiComp can be used in home environments (Rodden, Crabtree, Hemmings, Koleva, Humble, Akesson and Hansson 2004), environmental games (Flintham, Anastasi, Benford, Hemmings, Crabtree, Greenhalgh, Rodden, Tandavanitj, Adams and Row-Farr 2003), and learning through interaction with the environment (Rogers, Price, Randell, Stanton-Fraser, Weal and Fitzpatrick 2005).
In this paper, we discuss a case study in user-centred design of the latter form of UbiComp. We report on a project where we worked with an English heritage site, Chawton House (best known for its association with Jane Austen), to develop new kinds of educational experience for schoolchildren, one of the main visitor groups.

The curators of Chawton House were very interested in the potential of new technology to provide something new to visitors, but had no specific ideas, and no urgent problem to solve since they already had a working practice of showing people around on foot, telling them about the house and grounds. In contrast, the teachers we worked with (from a nearby school, Whitely Primary School) had a reasonably clear view of what they wanted to see offered by Chawton House: a technology-enhanced 'fieldtrip' to promote creative writing, a key skill in the literacy curriculum.

The project generated a number of issues and challenges that are arguably applicable to many designers of UbiComp. UbiComp is intrinsically novel, often involving arrangements of wireless infrastructures, computing components, handheld devices, location-sensitive information delivery and so on, all distributed and integrated with the environment to produce services and interactions that are relatively unfamiliar. This has two key implications. The first is that what can be done with UbiComp in terms of human activities is very open. Rather than engaging with 'problem spaces', where existing activities are analysed for breakdowns, bottlenecks, etc., which we then attempt to fix, UbiComp opens up 'opportunity spaces' where there may not be given problems to solve, but where there is great potential to do new things. The second implication of the novelty of UbiComp is that the process of engaging with users in user-centred design processes becomes more difficult as the technologies may be much less a part of their lives and existing practices: therefore, we are less able to assume that users already have an insight into how the technology works and what it is for. This issue is exacerbated by the complexity and situatedness of UbiComp which makes it more difficult to prototype and evaluate: frequently, UbiComp systems need to be implemented at at least hi-fi prototype level in the environment they were built for before they can be evaluated by users, with big implications for how we do prototyping at lo- and mid-fi levels - and indeed, how far this is meaningful.

In the rest of this paper we discuss how these issues impacted our co-design relationship with the key stakeholders in the project: the curators of Chawton House, and teachers from Whiteley Primary School, and how these were addressed.

2 The Co-Design Process

In this section we describe how our relationship with our co-design partners changed over time; how we mediated and managed the relationships between stakeholders; and how we scoped the opportunity space in partnership with the stakeholders. This is organised around discussion of our engagement with Chawton House and Whiteley School. For the purposes of this paper, we concentrate on six workshops we ran - three for each partner - together with discussion of a demonstrator fieldtrip which was designed and run as a result. More detail on these and other activities is given in (Halloran, Hornecker, Fitzpatrick, Weal, Millard, Michaelides, Cruickshank and De Roure 2006), and (Hornecker, Halloran, Fitzpatrick, Weal, Millard, Michaelides, Cruickshank and De Roure 2006). Fig. 1 shows when and where the workshops and fieldtrip occurred; images from the workshops and fieldtrip appear in Fig. 2 (over).

Date	Location	Activity
19.04.05	Chawton House	First curator workshop
20.04.05	University of Southampton	First teacher workshop
03.05.05	Chawton House	Second curator workshop
18.05.05	Whiteley School	Second teacher workshop
26.05.05	Chawton House	Third curator workshop
03.06.05	Chawton House	Third teacher workshop
12.07.05	Chawton House	Fieldtrip

Fig. 1. Workshops - schedule.

2.1 Workshops

The first curator workshop had three aims: to understand curators' current practices, to find out what kinds of things they already tell visitors about the grounds, and to discuss possible sorts of tours for visitors. We printed a large map of Chawton House including its gardens, and populated it with models of buildings to ground discussion. The workshop provided us with an initial understanding of curators' practices, and provided curators with a beginning understanding of our design vision. However, this workshop also raised a number of issues in terms of building a relationship, scoping the space of possibilities, and mediating between Chawton House and Whiteley School. Originally we had wanted to record some of the stories visitors are told, for possible use in a system, but we found that curators were not used to telling stories when not on location. This reflects that at the beginning of the design relationship we made assumptions about practice that we needed to revise, an example of us needing to learn from them to understand their practice. Conversely, our ideas about the system were necessarily unspecified.

During the first workshop with teachers we designed a rough structure for a fieldtrip, using the same map, to help teachers remember the features of the grounds (which they had visited earlier). The workshop gave us insight into how teachers design fieldtrips, their educational value, and how they are organized. The map focused discussion about the event's structure and general orchestration.

It became apparent at this point that the co-design challenges were not uniform across the two sets of stakeholders, i.e. the curators of Chawton House and the teachers. The space of possibilities for the curators, around new ways of giving tours, was much less constrained than for the teachers, who were specifically working with how to introduce technology into an existing practice, fieldtrips. The relationship with each group would thus be different. For future workshops we needed to manage the sequencing of workshops, processing the results of each to inform subsequent workshops, and mediating between co-design partners, so that the three stakeholders (ourselves, the curators, and the teachers) could work together effectively.

During the second workshop with curators, each of the three curators we worked with - Alan, Sue and Greg (names have been changed) - took a pair of researchers on separate guided tours, which we videotaped. Through this experience, together with our prior observations of another tour, as well as talking to the curators, we came to see how the curators' creation of visitor experiences of Chawton House is a skilled, dynamic, situated and responsive activity, a form of improvisation triggered by locations, artefacts and visitors' responses and questions. This revealed more about how tours are actually conducted than the map activity in the first curator workshop.

An aim of this workshop was to elicit the 'content' for possible use in a UbiComp system that we had been unable to effectively elicit during the first curator workshop. Following review and analysis of the recordings, we were struck by the authority, humour and energy of the curators' talk, and made a provisional decision to use these 'authentic' audio segments for a system, given curators' agreement.

Fig. 2. Workshops. Clockwise from top left: working around a 'tangible map' at the first curator workshop; researchers being given a tour of the grounds during the second curator workshop; children interacting with part of the system during the fieldtrip; teachers working around the tangible map at the first teacher workshop.

In the second teacher workshop, we had a set of audio clips that the teachers could use to develop their ideas. It was decided to select clips that provided historical or social information that children could listen to, and to use these in conjunction with instructions and prompts from teachers, that the system could display. The teachers used the map to place notes where events could happen and instructions given. However, when it came to deciding on concrete activities and instructions, the teachers hesitated, as we still needed to establish exactly what was possible. To promote teachers' understanding of the possibilities, we described, and showed pictures of, a set of technologies from another project that were a candidate for this one: mobile devices with audio and text, capable of sensing location. Thus, this workshop implied increased commitment to an already existing, candidate suite of technologies.

The third curator workshop was aimed at increasing curators' understanding of the technology. We presented an example of a related system for school fieldtrips (used

on another of our projects) by means of a video (the same system as explained to the teachers). In addition, we toured the grounds playing selected audio clips in different locations from a laptop to give an impression of how the children might experience these. On this basis, the curators agreed that these clips could be used in the system.
An important output in terms of our developing relationship with the curators was the level of ownership of the system. We came to understand that curators were not yet sure what value the system could provide them with and thus were hesitant to invest effort. Although they were interested in our feedback on their practices, stating that it was "interesting to see what you pulled out [the audio clips], what you find interesting", the workshop revealed that the devised system was still seen as designed and 'owned' by researchers, indicated by Sue saying: "once you've decided what you want to include" [our italics]. An important implication of this was that we needed to increase curators' engagement with the fieldtrip.
The third workshop with teachers took place at Chawton House. We used this opportunity to introduce the teachers to the curator who would give a tour of the house on the day of he fieldtrip. Then we walked the grounds, the teachers brainstorming ideas for activities and instructions, assisted by us with background information and an overview of suitable audio clips. Back in the house, ideas were selected and refined and the timings planned for e.g. how long children should stay at a location and how instructions would be sequenced. Further collaboration (via email) consisted of sharing notes, writing instructions, and refining the orchestration. The third teacher workshop, then, was focussed and bounded by teachers' direct experience of the location and what we said was possible in terms of the technology.

2.2 The System

During the workshop process and given the willingness of both sets of stakeholders to agree to our ideas about what technology was possible, we developed a system consisting of portable devices capable of delivering and recording audio and text. These devices, an arbitrary number of which could all be used at the same time, were linked to a location-sensing architecture consisting of GPS augmented by pingers (RF beacons). The content (audio clips, text instructions) was organised and delivered by means of an information architecture based on adaptive, physical hypertext, which is sensitive to prior locations and content already received. Users could record audio and text messages ('annotations'). The system logged movements and annotations and the results could be accessed on the PDA and later by users on web logs. A fuller technical description of the system can be found in (Weal, Cruickshank, Michaelides, Millard, De Roure, Hornecker, Halloran and Fitzpatrick 2006).

2.3 The Fieldtrip

The two-hour school fieldtrip took place four months after the project began. We invited curators to observe it, to provide them with direct experience of the system in use. Two curators were present on the day, observing and following the children. Six children, as well as the two teachers, came. First, Sue gave the children a guided tour of the house. Then the children explored the grounds in pairs, free to go wherever they wanted, and followed by researchers recording them. Each pair shared a portable device with location sensing, which could also record audio and text. The device introduced the children to a location with audio clips. It then displayed a series of prompts designed to inspire children's imagination. For example, after listening to a clip about a location called The Wilderness, they were asked to explain the reason for this name in their own words, and instructed to find a 'spooky' spot and describe it. After this phase, children met with the teachers, decided on initial ideas for a story and two locations to focus on. Then they went to these places and were prompted by the system to conceptualize a story. The next day at school the children continued writing their stories. A fuller description and analysis of the fieldtrip can be found in (Halloran, Hornecker, Fitzpatrick, Weal, Millard, Michaelides, Cruickshank and De Roure [2] 2006).

3 Discussion

Promoting users' understanding of UbiComp technologies is a challenge, not only due to the novelty of the technology, but also because it allows for the creation of novel practices. With mobile and distributed systems, it is very difficult to provide an adequate idea of how the system will work until it has been built. On the other hand, we did not wish to pre-empt the co-design process by presenting a system as a fait accompli. Prior to the fieldtrip, we tried a number of techniques to overcome this problem, but none was ideal. Showing videos of related systems did not provide actual experience and there were differences in application from their context. Walking around with a laptop to play clips in-situ had been partly successful. This had already required authoring of content and postproduction of clips. Playing clips on the actual device in the right order required large parts of the data defined and the infrastructure in place. Allowing users to experience the technology and from this to envision further options with UbiComp often means that researchers need to invest significant effort. This is an ongoing issue in co-design of UbiComp: the practice of creating meaningful prototypes at lo- and mid-fi levels from simple, cheap and disposable materials is much less available, and this has a big impact on the use of prototyping as a tool for discussion and development.

This said, the fieldtrip provided stakeholders with a much clearer vision of what the new technology could provide them than any of our prior attempts. Sue told us in an interview directly after the fieldtrip: "It was nice to be able to see the system working. Not being technically minded, it didn't mean a great deal to me to begin with". Curators could see the fieldtrip as a template for other creative writing activities for a diverse range of visitors; they liked how visitors would be able to control the pace and order of a tour. Teachers similarly told us that "we were not quite sure about the technologies. And now we've seen them, we've got a much better understanding". While designing the fieldtrip, they had at times been worried whether "what you're writing down, would that actually work" and would it get the best out of the technology.

Our experience confirms that users' acquiring an understanding of UbiComp, and envisioning new practices requires time and cannot be rushed. For the curators, for example, 'springboards' or 'disturbances' for rethinking were provided predominantly by the fieldtrip, and to some extent also through the audio segments and transcripts presented earlier. Readiness and openness to being exposed to these disturbances relied on the evolution of a design relationship.

Another important factor was mutual learning. From the curators, we learned a lot about the estate and their work practices, understanding what they care about. From teachers we learned about practices around fieldtrips. The mutual learning taking place, which at times meant that our design partners had to educate us how to approach things best (like Alan in the first workshop telling us to tape actual tours), is reflected by Alan ironically saying there had been "a learning curve for everybody" in an interview. Again, in co-design of novel UbiComp scenarios, this may be inevitable, but requires commitment from all stakeholders.
Willingness to be open to 'disturbances', and to engage in learning, had an important influence on the curators' input into the co-design process. In our data we can find several examples of how the engagement with us inspired them to rethink their practices. For example, Greg reflected "one of the real attractions of this thing [the system] (...) the gardens are best experienced in solo or very small groups, whereas the house, it does not matter so much (...). The open spaces, there's a different feeling, where a more intimate personal approach is good... You might with a machine get a more personal approach, which is just you and the machine, rather than you and 14 others and a guide". This demonstrates how curators became increasingly open to the idea of a technological rather than personal guide system and their imaginations were stimulated.

4 Conclusion

In this paper we have described the issues that arise from engaging with users to co-design UbiComp technologies, in order to augment and enhance their practices and activities, without any given urgent problem. Success factors included iteration through a diversity of design activities; providing hands-on experience and a concrete example of a visitor experience that was novel, and emerged out of mutually developing understandings over time; mediating between stakeholders, slowly bringing them into a more direct relationship with each other; and, finally not being 'distanced' (objective) researchers, but truly engaging with the setting and caring about the same things as our co-design partners.
The key challenge of this research was how to create a meaningful co-design relationship with stakeholders where people are time-pressured and the engagement is about re-envisioning and creatively imagining new things rather than solving present problems. Perhaps inevitably, such agendas are not likely to be top priority. However, this does not reduce the urgency of this type of initiative. New technologies offer novel and even radical new ways of delivering value to users, and techniques of engaging with users need to be developed in offer to deliver this value despite pressured contexts, in order to realize novelty rather than recreating what is already known. Here, we have started to investigate what is involved. To take co-design seriously in opportunity spaces, we have to be reflective practitioners who carefully and continuously promote the value of user involvement to get progressive buy-in, against a background of developing understandings of user needs and practices and what is meaningful to them as these develop during the design process.

Acknowledgements

This research was funded by the EPSRC IRC project 'EQUATOR', GR/N15986/01. The authors thank Chawton House and Whiteley Primary School; Geraldine

Fitzpatrick, Mark Stringer, Rowanne Fleck, Eric Harris, Anthony Phillips, and Anna Lloyd at the University of Sussex; and Mark Weal, David Millard, Danius Michaelides, Dan Cruickshank, and Dave De Roure at the University of Southampton.

References

Flintham, M., Anastasi, R., Benford, S., Hemmings, T., Crabtree, A., Greenhalgh, C., Rodden, T., Tandavanitj, N., Adams, M. and Row-Farr, J. (2003) Where on-line meets on the streets: Experiences with mobile mixed reality games. *Proc. CHI '03*, 569-576.

Halloran, J., Hornecker, E., Fitzpatrick, G., Weal, M., Millard, D., Michaelides, D., Cruickshank, D. and De Roure, D. (2006) Unfolding understandings: Co-designing UbiComp in situ, over time. *Proc. DIS 2006*, 109-118.

Halloran, J., Hornecker, E., Fitzpatrick, G., Weal, M., Millard, D., Michaelides, D., Cruickshank, D. and De Roure, D. [2] (2006) The literacy fieldtrip: Using UbiComp to support children's creative writing. *Proc. IDC 2006*, 17-24.

Hornecker, E., Halloran, J., Fitzpatrick, G., Weal, M., Millard, D., Michaelides, D., Cruikshank, D. and De Roure, D. (2006) UbiComp in opportunity spaces: Challenges for participatory design. *Proc. PDC 2006*, 47-56.

Rodden, T., Crabtree, A., Hemmings, T., Koleva, B., Humble, J., Akesson, K-P., and Hansson, P. (2004). Between the dazzle of a new building and its eventual corps: Assembling the ubiquitous home. *Proc. DIS 2004*, 71-80.

Rogers, Y., Price, S., Randell, C., Stanton-Fraser, D., Weal, M., and Fitzpatrick. G., (2005) Ubi-learning: Integrating outdoor and indoor learning experiences. *CACM, 48(1)*, 55-59.

Weal, M., Cruickshank, D., Michaelides, D., Millard, D., De Roure, D., Hornecker, E., Halloran, J. and Fitzpatrick, G. (2006) A reusable, extensible infrastructure for augm field trips. *Proc. PerCom 2006*, 201-205.

Weiser, M. (1991) The computer for the twenty-first century. *Scientific American*, 265(3 104.

Section 5

Semantic Information Retrieval Workshop

Association Analysis of Alumni Giving: A Formal Concept Analysis

Ray R. Hashemi[1], *Louis A. Le Blanc*[2], *Mahmood Bahar*[3], *and Bryan Traywick*[1]

1 Armstrong Atlantic University, Department of Computer Science, Savannah, GA 31419, USA, hashemra@mail.armstrong.edu
2 Berry College, Campbell School of Business, Mount Berry, GA 30149-5024, USA, laleblanc78@comcast.net
3 Teachers Training University, Department of Physics, Tehran, Iran

Abstract. A large sample (initially 33,000 cases representing a ten percent trial) of university alumni giving records for a large public university in the southwestern United States are analyzed by Formal Concept Analysis (FCA). This likely represents the initially attempt to perform analysis of such data by means of a machine learning technique. The variables employed include the gift amount to the university foundation (UF) as well as traditional demographic variables such as year of graduation, gender, ethnicity, marital status, etc.
The UF serves as one of the institution's non-profit, fund-raising organizations. It pursues substantial gifts that are designated for the educational or leadership programs of the giver's choice. Although they process gifts of all sizes, the UF focus is on major gifts and endowments.
The Association Analysis (AA) of the given dataset is a two-step process. In the first step, the data items that are frequently appear together (i.e. *concepts*) are systematically identified and in the second step, each concept is converted into a set of rules called *association rules*. The hypothesis examined in this paper is that the generosity of alumni toward his/her alma mater can be predicted using association rules obtained by applying the FCA approach.

1 Introduction

Charitable and philanthropic organizations, including university fund raising departments, face increasing pressure to more effectively employ a variety of analytical techniques (Brown, 2004). Potential donor identification may provide one such tool (Shelley and Polonsky, 2002). As applied in other public sector or not-for-profit agencies (Johnson and Garbarino, 2001; Todd and Lawson, 2001; Wymer, 2003), segmentation can assist college and university development officials in identifying potential donors and help define what best to communicate to potential donors.
Research studies report various methodologies and techniques to identify giving behavior for collegiate financial development. Willemain et al. (1994) employed a

general linear model to predict university alumni giving. Lindahl and Winship (1992) used logit analysis to predict rare events such as gifts over $100,000. Clotfelter (2001) utilized basic descriptive statistics (e.g., means, percentages) to portray survey results from two generations of alumni giving.

With somewhat more novel quantitative approaches, Drye et al. (2001) suggested survival analysis (based upon logarithmic charts) to better identify the most regular supporters and those most likely to repeat their support. Key (2001) recommended probit regression for building a response model to identify individuals most likely to make a major, capital or planned gift.

The research application reported herein recognizes that neither money nor time is available to solicit directly from every member an alumni base of nearly a quarter million people. Some method needs to be devised to identify those individuals who are very likely to give as well as those alumni with significantly smaller probabilities of charitable giving. In this research effort, we examine the hypothesis that the generosity of alumni toward his/her alma mater can be predicted using association rules obtained by using the FCA approach.

The organization of the remainder of this paper is as follows. In Section 2.0, Formal Concept Analysis (FCA) is described; and, in Section 3.0, the methodology is presented. Section 4.0 covers the empirical results. Section 5.0 briefly contains the conclusion and ideas for future research.

2 Formal Concept Analysis

The FCA is an approach introduced by R. Wille (1982) for identifying the formal concepts of a given dataset. The details of this approach may be found in Deogun et al. (1998) and Ganter and Willie (1999). A formal concept of a dataset is identifiable based on the formal context of the dataset.

A *formal context* K is a triplet (G, I, A) in which G is a set of objects, A is a set of attributes, and I is a relation between G and A (i.e. $I \subseteq G \times A$). For context K=(G, I, A), let $H \subseteq G$ and $B \subseteq A$. We define two operators of H' and B' as follows:

$H' = \{a \in A \mid gIa \text{ for all } g \in H\}$ and $B' = \{g \in G \mid gIa \text{ for all } a \in B\}$

A *formal concept* of the context K= (G, I, A) is a pair (H, B) such that $H \subseteq G$, $B \subseteq A$, H' = B, and B' = H. Let C1 = (H1, B1) and C2 = (H2, B2) be two concepts of the context K = (G, I, A). If $H1 \subseteq H2$ (this implies that $B2 \subseteq B1$), then C1 is a *sub-concept* of C2 and C2 is a super-concept of C1.

If we use the relation $\leq$ as a hierarchical order, then we can write $C1 \leq C2$. If all the concepts of the context K be ordered using the $\leq$ order, then the result is a *concept lattice* denoted by L(G, I, A). For two concepts C_i and C_j, $(i \neq j)$ the concepts $(C_i \wedge C_j)$ and $(C_i \vee C_j)$ are the *infimum* (also called *meet*) and *supremum* (also called *join*) concepts of C_i and C_j, respectively.

3 Methodology

For a given dataset, D, a *concept,* in FCA nomenclature, is a set of data items that frequently appear together. And it is the same as the *frequent itemset* commonly used in data mining discipline (Han and Kamber, 2005). If the number of records that contain all the data items of a given concept, C_i, is N_c, then C_i has the support value of S_{Ci} and it is defined as $S_{Ci} = N_{Ci} / |D|$.

The concept C_i can be easily converted into a set of *association rules*. Each association rule in the set has the general form of $A_i \rightarrow B_i$ where, A_i is the conditions set and B_i is the conclusions set. The rule $A_i \rightarrow B_i$ obeys the following constraints:

- $A_i \subset C_i$, $A_i \neq \emptyset$, $B_i \subset C_i$, $B_i \neq \emptyset$,
- $A_i \cup B_i = C_i$, $A_i \cap B_i = \emptyset$,
- All data items in A_i are logically anded together as are those in B_i.

If $|B_i| = 1$, then the association rule $A_i \rightarrow B_i$ is *simple*. Considering the above constraints, the total number of association rules that can be generated from a concept with n data items is $2^n - 2$ from which n of them are simple.
Each association rule is assigned a *support* and a *confidence*. The support for the rule $A_i \rightarrow B_i$ that is extracted from concept C_i is equal to S_{Ci} and the confidence, K, is defined as $K_{(Ai \rightarrow Bi)} = S_{Ci} / S_{Ai} = N_{Ci} / N_{Ai}$.
Example, The dataset D is given. Let the non-simple association rule of "a&b → c&d" be one of the association rules generated from the concept $C_i = \{a, b, c, d\}$. Let also the $N_{Ci} = 20$, $N_{a\&b}=30$, and $|D| =200$. The support, $S_{Ci}= 20/200 = 0.1$ says that in only 10 percent of the records in D, data items of a, b, c, and d collectively appear. The confidence, K=(20/30=0.64) says that 64 percent of the records in D that contain data items a and b also contain data items c and d.

3.1 Reduction of Concepts and Association Rules

The application of FCA on a dataset generates a large number of concepts. The sheer volume of the concepts makes the use of them impractical. To ease this problem, first we introduce a new terminology and then present a set of criteria for pruning the concepts.
Suppose a dataset about the alumni of a university is given and we intent to analyze the alumni contributions to the university. Let us assume that one of the attributes in this dataset is the "Contribution Amount" with possible three categorical data items of zero, one and two that mean "No Contribution", "Small Contribution" and "Large Contribution", respectively. Obviously, we are only interested in those concepts that contain one of these three data items. In our analysis, we refer to these categorical data items as *designated* data items.

The pruning of the concepts is done based on the following two criteria:

1. Concepts that do not include a *designated* data item are removed.
2. Concepts with support less than threshold T1 are removed; and,

The algorithm FilterConcepts uses the above criteria and delivers pruned concepts.

3.2 Algorithm FilterConcepts

Given: A set of concepts, χ, generated using FCA. Each concept, C_i, has a support, S_{Ci}. The designated set of data items, V. The threshold value of T1.
Objective: Pruning the concepts.

Step 1. Repeat Step 2 for i = 1 to $|\chi|$;
Step 2. If $((S_{Ci} < T1) \,||\, (C_i \cap V) = \emptyset))$ then remove C_i from χ;
Step 3. End;

Upon the completion of the concepts pruning, the association rules for every concept is created and filtered using the following criteria:

1. All non-simple rules are removed,
2. All simple rules that their "conclusion" sections are not one of the designated values are removed, and
3. All simple rules with confidence less than a threshold value of T2 are removed.

The algorithm FilterRules uses the above criteria to deliver the simple rules of interest.

3.3 Algorithm FilterRules

Given: A set of association rules, R, generated from the filtered set of concepts. The threshold values of T2. The confidence, K_i, for the rule r_i.

Objective: Pruning the association rules.

Step 1. Repeat Steps 2 for i = 1 to |R|;

Step 2. If ((K_i < T2) || (r_i.conclusion ∩ V) = ∅) then remove r_i from R;

Step 3. End;

The number of association rules is further reduced by collapsing the associated rules using the following criterion:

If two rules, r_i and r_j, have the same conclusions and
the common data items in their conditions sections is greater than threshold value T3 and
the number of non-common data items in their conditions section is less than threshold value T4

Then If $|r_i| \le |r_j|$
Then remove r_j from the pruned rule set;
Else remove r_i from the pruned rule set;

The algorithm Collapse delivers the outcome of applying the above criterion on the pruned set of association rules.

3.4 Algorithm Collapse

Given: A set of pruned association rules, R'. The threshold values of T3 and T4.

Objective: Collapsing the association rules.

Step 1. Repeat Steps 2 to 3 for i = 1 to |R'|;

Step 2. Repeat Steps 3 for j = i+1 to |R'|;

Step 3. If (r_i.conclusion = r_j.conclusion) &((r_i.condition ∩ r_j.condition) ≥ T3) &
(r_i.condition - (r_i.condition ∩ r_j.condition) ≤ T4) &
(r_j.condition - (r_i.condition ∩ r_j.condition) ≤ T4) &
Then If (|r_i.condition| ≤ - |r_j.condition|)
Then remove r_j from the pruned rule set; K_i = Min(K_i, K_j);
Else remove r_i from the pruned rule set; K_j = Min(K_i, K_j);

Step 4. End;

The resulting collapsed association rules are grouped based on the values of their conclusions. Since the conclusions are the same as the designated values, each group is a set of association rules supporting one of the designated values.

4 Empirical Results

A dataset with 32,901 records that carry information about alumni of a well known university is obtained. Due to overwhelming number of missing data, use of any data imputation approach may generate artificial concepts. Therefore, we extract the association rules from only those records with no missing data and then validate the association rules by using the remaining of the data. Since a concept is made up of a subset of data items, then there is a chance that it matches a number of test records.
The number of records with no missing data is 774. Each record is composed of eight attributes, Table 1. The attribute "GiftAmount" is the designated attribute and, therefore, its data items are designated values. The designated values are 0, 1, and 2 that stand for "$0", "Greater than $0 and less than $100,000", and "Greater than or equal to $100,000" respectively.
The FCA approach is applied to the test set and the resulting concepts are pruned, their association rules are generated, pruned, collapsed, and separated for designated data items using Algorithms FilterConcepts, FilterRules, and Collapse with threshold values of T1=0.02, T2=0.6, T3=4, and T4=1.

Attribute Name	Attribute Definition	Number of Categorical data Items
ClassYear	Alumni's Year of Graduation	10
School	Alumni's Granting Degree School	6
Gender	Alumni's Gender	2
Etnicity	Alumni's Ethnicity	2
MaritalStatus	Alumni's Marital Status	2
CorpFlag	Alumni's Service in the Student Corp of Cadets	2
WelthIndex	Alumni's Index of Wealth	4
GiftAmount	Alumni's Contribution to the Foundation	3

Table 1: List of Attributes.

Before conducting the test, we have concluded that those alumni with a wealth index less than 3 are not in a financial situation that can make a contribution. Therefore, we pruned the association rules even further by removing those rules which carry a wealth index less than three. The numbers of rules in the final set for the designated values of GiftAmount = 0, 1, and 2 are 5, 75, and 1, respectively.

Measures	Using only Association rules with conclusion = GiftAmount of zero	Using only Association rules with conclusion = GiftAmount of one	Using the combination of both set of Association rules
No. of hits	204	4511	4715
No. of true pos.	162	3367	3367
No. of true neg.	0	0	162
No. of false pos.	42 (21%)	1144 (25%)	1144 (24%)
No. of false neg	0 (0%)	0 (0%)	42 (0.8%)
Prediction Quality	79.4 %	74.6 %	74.8 %

Table 2: The result of statistical measurements of the test set.

The association rules are applied on the test set of (32,901-774) 32,127 records. The total number of hits is 4720. However, the number of hits for records with GiftAmount = 2 is 5. As a result, we have removed those 5 records from the test set and the number of hits reduced to 4,715. The quality of prediction and the numbers of false positive (F+), true positive (T+), false negative (F-), and true negative (T+) for GiftAmount = 0, and 1 are calculated separately and in combination. The results are illustrated in Table 2.

5 Conclusion and Future Research

The results of Table 2 reveal that the hypothesis is true. Association Analysis of the alumni data generated association rules that are able to predict the generosity of the alumni to the university and on average, the quality of the prediction is 74.8 percent with sensitivity of 98.7 percent and specificity of 12.5 percent. Considering the facts that (a) the association rules concluded from only 774 records, less than 2.4% of the total records and (b) the actual test set (4715/774) that is 610% greater than the training set, the results displayed in Table 1 are quite significant.

The creation of profiles for alumni who are "generous" and alumni who are "not so generous" represent a future line of research that has already commenced. These profiles would help the UF target the potentially "generous" graduates for contribution to the university.

References

Brown, D. W., (2004) "What Research Tells Us about Planned Giving," *International Journal of Nonprofit and Voluntary Sector Marketing*, 9(1): 86-95.

Clotfelter, C.T. (2001) "Who Are the Alumni Donors? Giving by Two Generations of Alumni from Selective Colleges," *Nonprofit Management & Leadership*, 2(2): 119-138.

Deogun, J. S., J. S. Raghavan H. and Sever (1998) "Association Mining and Formal Concept Analysis, *Proceedings of the Joint Conference in Information Science*, 335-338.

Drye, T., G. Wetherill and Pinnock, A. (2001) "Donor Survival Analysis: An Alternative Perspective on Lifecycle Modelling," *International Journal of Nonprofit and Voluntary Sector Marketing*, 6(4): 325-334.

Ganter, B. and R. Wille (1999), *Formal Concept Analysis: Mathematical Foundations*, Berlin: Springer-Verlag.

Han, J. and M. Kamber (2005) *Data mining, Concepts and Techniques*, Morgan Kaufmann Publishers, 2nd Edition.

Johnson, M.S. and Garbarino, E. (2001) "Customers of Performing Arts Organizations: Are Subscribers Different from Nonsubscribers?" *International Journal of Nonprofit and Voluntary Sector Marketing*, 6(1): 61-77.

Key, J. (2001) "Enhancing Fundraising Success with Custom Data Modeling," *International Journal of Nonprofit and Voluntary Sector Marketing*, 6(4): 335-346.

Lindahl, W.E. and Winship, C. (1992) "Predictive Models for Raising and Major Gift Fundraising," *Nonprofit Management & Leadership*, 3(1): 43-64.

Shelley, L. and Polonsky, M.J. (2002) "Do Charitable Causes Need to Segment Their Current Donor Base on Demographic Factors?: An Australian Examination," *International Journal of Nonprofit and Voluntary Sector Marketing*, 7(1): 19-29.

Todd, S. and Lawson, R. (2001) "Lifestyle Segmentation and Museum/Gallery Visiting Behaviour," *International Journal of Nonprofit and Voluntary Sector Marketing*, 6(4): 269-277.

Wille, R. (1982) "Restructuring Lattice Theory: An Approach Based on Hierarchies of Concepts," in: *Ordered Sets*, I. Rivali, ed., Reidel Dordecht Publisher, Boston, 1982, 445-470.

Willemain, T. R., A. Goyal, M. van Deven and Thukral, I.S. (1994) "Alumni Giving: The Influences of Reunion, Class and Year," *Research in Higher Education*, 35: 609-629

Wymer, W.W. Jr.(2003) "Differentiating Literacy Volunteers: A Segmentation Analysis for Target Marketing," *International Journal of Nonprofit and Voluntary Sector Marketing*, 8(3): 267-285.

Comparison of Automatic Clustering and Manual Categorization of Documents

Kazem Taghva and Meghna Sharma[1]

1 University of Nevada, Las Vegas, Information Science Research Institute,
taghva@isri.unlv.edu.edu, sharma@isri.unlv.edu

Abstract. The fundamental goal of this research is to learn whether unsupervised learning can be used to cluster documents in the collection in a similar way that manual categories are. We report on our experiments with K-mean clustering algorithm to provide a partial answer to the above mentioned goal.

1 Introduction

Data mining is typically used to extract new facts and relationships from data. Some of the obvious applications of data mining include extraction of relationships between various attributes of structured data. For example, concluding that in the retail environment customers that buy certain items are more likely to buy other related items is a simple application of data mining. Data Mining can also be used to extract and relate pieces of information in textual documents. Discovery of acronyms and their definitions is an example of this kind of data mining (Taghva and Gilbreth 1995).

One of the most commonly used data mining techniques is document clustering or unsupervised document classification. Clustering techniques are used to classify documents to form clusters which are similar in content. Document objects can be clustered either based on distance, which means if the data objects are within close proximity (decided by a proximity function) they are grouped together in one cluster, or based on a common concept where the data are clustered together if they fit a common conceptual description (Wikipedia 2006).

Another well-known data mining technique is categorization or supervised learning. Categorization techniques learn from manually categorized documents to classify future incoming documents, i.e., a set of training data (manually categorized documents) are used to aid categorization of new documents.

An interesting question regarding clustering and categorization is whether there is a link between these two techniques. The intention of manual categorization is to form groups of similar documents. Automatic categorization then learns from these manually categorized documents and places future incoming documents into proper groups.

Clustering is supposed to put the documents in various groups based on document similarity. Is it reasonable to expect some clusters to be fully contained in some category or vice versa? The objective of this research is to at least provide a partial answer to this question.
The remainder of this paper is divided into 3 sections. Section 2 briefly describes simple techniques in categorization and clustering. Section 3 is our experimental attempt at answering the question above. The final Section provides our conclusion and future work.

2 Background

There are many statistical-based techniques for text categorization of which the simplest, introduced in (McCallum 1996), is based on the multinomial naive Bayes model{McCallum98}. Following McCallum and Nigam (McCallum and Nigam 1998), assume we have a vocabulary $V = (w_1, w_2, \ldots, w_{|V|})$ for our collection, then a document d_i can be represented by a vector:

$$d_i = (N_{i1}, N_{i2}, \ldots, N_{i|V|}) \qquad (1)$$

where N_{ij} is the number of occurrences of the word w_j in the document d_i. We also assume we have a set of $C = \{c_1, c_2, \ldots, c_{|C|}\}$ classes that we want to assign to our document collection. One basic assumption is that each document falls into exactly one category (i.e., exhaustive and incompatible).
In this framework, we are interested in finding $P(c_j \mid d_i)$, or the conditional probability that a document belongs to category c_j. Using Bayes' theorem, we can calculate this probability by:

$$P(c_j| d_i) = P(c_j) * (P(d_i|c_j) / P(d_i)) \qquad (2)$$

In other words, Bayes' theorem provides a method to compute $P(c_j|d_i)$ by estimating the conditional probability of seeing particular documents of class c_j and the unconditional probability of seeing a document of each class. If we make the word independence assumption which states that the probability of each word occurring in a document is independent of the occurrences of other words in the document, then this probability can be estimated by:

$$P(d_i \mid c_j) = P(|d_i|)|d_i|! \, \Pi_t P(w_t|c_j)^{N_{it}} / N_{it}! \qquad (3) \qquad t \text{ ranges from 1 to } |V|$$

In this formula, the $P(w_t|c_j)$ is computed using word frequencies in training documents. Assuming a training set of documents $D = \{ d_1, d_2, \ldots, d_{|D|}\}$ and the fact that we have an exhaustive and incompatible set of classes, then

$$P(w_t|c_j) = 1 + \Sigma_i N_{it} P(c_j|d_i)/|V| + \Sigma_s \Sigma_i N_{is} P(c_j|d_i) \qquad (4)$$

i ranges from 1 to $|D|$ and s ranges from 1 to $|V|$

In general, words are stemmed and stop words are removed to decrease the size of the vocabulary. In addition, various dimensionality techniques are used to also reduce the vocabulary size. In our experiments, we generally remove the words with high and low document frequency.
There are also many known techniques for clustering of which the simplest is the k-means algorithm (Kanungo et al. 2002; Weisstein 2006). The total number of

clusters that are produced is predefined. So the algorithm takes data points and the value k, which refers to the number of clusters the points must be grouped into, as inputs and groups them into k clusters where the clusters are recognized by a cluster centroid. All the data points belonging to one cluster are closer to the centroid of that particular cluster than any other cluster's centroid. The main idea behind the algorithm (Kanungo et al. 2002) is to minimize the objective function which is shown below:

$$J = \sum_j \sum_{x,u} | x_n - u_n | \qquad (5)$$

where u_n refers to the nth data point of the centroid indexed by j (j ranges from 1 to k), x_n refers to the nth data point of the document we are clustering. Basically, this equation says that the clusters are formed by minimizing the distance between the data point and the centroid.[3]

The algorithm starts with picking k random points as centroids for the clusters. In the next stage, the algorithm assigns each point from the data set to a centroid which is closest to it. Once all the points have been assigned to a centroid, the first step of the algorithm is complete. In the next step, the algorithm recalculates the centroids of each cluster. This is done by finding the center of the clusters produced in the first step. The center should be a point which is equidistant from every other point in that particular cluster. The next step is to re-assign every point in the data set to a centroid/cluster by calculating the distance between every point and every cluster and choosing the one with the minimum distance. Once this is done, new centroids are calculated. This loop repeats itself until the k centroids do not change their location.

The above k-means clustering algorithm is a general purpose algorithm.

In our setting, we need to adapt it to document clustering. As in the case of categorization, we reduce the size of the dictionary in order to reduce the vector dimensions. In addition, the nth data point for each document is calculated according to the standard TF*IDF weighting algorithm.

Categories	No. of Documents: Reuters 213	No. of Documents: Reuters 324
MCAT	558	578
CCAT	1618	*1356*
ECAT	*375*	*256*
GCAT	*553*	507
Total Docs:	3104	2697

Table 1. Reuters sub-collections

3 Experimental Results

We set up a small experiment to establish a preliminary result. We decided to run our experiments on a small group of documents from the Reuters Corpus, Volume 1, English language, 1996-08-20 to 1997-08-19 (Release date 2000-11-03, Format

[3] The objective function in this case is the squared error distance between the data point and the centroid of the cluster.

version 1, correction level 0), called Reuters213 and Reuters324 (University of California 2006) with the following four categories:
MCAT - Markets
CCAT - Corporate/Industrial
ECAT - Economics
GCAT - Government/Social
A description of the Reuters sub-collections appears in Table 1. These documents are first scanned for stop word removal. The next step was stemming using Porter's Stemming Algorithm (Rijsbergen, Robertson and Porter 1980). In addition to stopword removal and stemming, we changed the standard Euclidean distance similarity function to a cosine measure as it is used for document similarity.

Cluster Number	No. of Documents
0	97
1	102
2	493
3	67
4	246
5	435
6	54
7	159
8	900
9	144

Table 2. Cluster runs

In general, the number of clusters, the initial location of centroids and the distance function can affect clustering. We chose only to vary the number of clusters. We run the k-means for k values of 5, 10, 20, and 30. The run of 10 clusters is a typical experiment and tabulated in Table 2. In this experiment, we observed that some clusters have most of their documents from the same category. For example, cluster 7 has all its documents from the same category CCAT - Corporate/Industrial. We also observed that documents from two different categories are clustered together due to common words that are not category specific. By doing further dimensionality reduction, we can rid clusters of some of these common words. Table 3 summarizes our runs for 20 clusters using this expanded dimensionality reduction.By analyzing this data, we observed that many of these clusters have a majority of their documents from the same category. For example, cluster 0 and 3 contain documents from the same category CCAT – Corporate/Industrial.

Cluster	**No.of DocsMCAT**	**No.of DocsGCAT**	**No.of DocsCCAT**	**No.of DocsECAT**	**Total No. of Docs**
0	0	0	83	0	83
1	17	0	69	64	150
2	21	0	16	18	55
3	0	0	41	0	41
4	0	2	289	0	291
5	0	83	1	0	84

6	39	0	62	3	104
7	1	28	116	40	185
8	0	247	71	51	369
9	9	92	29	5	135
10	33	0	51	48	132
11	7	33	37	15	92
12	0	57	135	9	201
13	119	3	59	113	294
14	7	3	22	3	35
15	0	2	55	0	57
16	111	2	209	1	323
17	40	0	12	3	55
18	154	1	63	1	219
19	0	0	198	1	199

Table 3. Cluster runs with dimensionality reduction

4 Conclusion and Future Work

As mentioned, the fundamental goal of this research was to learn whether unsupervised learning can be used to cluster documents in the collection in a similar way as supervised categorization. This preliminary work shows that there is definitely a relationship between clusters and categories. This work also shows that due to the nature of free text, it may be hard to find the right k such that all documents from the same cluster belong to the same category.

References

Kanungo T., Mount D. M., Netanyahu N., Piatko C., Silverman R. and Wu A. Y. (2002) An Efficient k-Means Clustering Algorithm: Analysis and Implementation. In: IEEE Trans. PAMI vol. 24, pp. 881-892.

McCallum A. (1996) Bow: A Toolkit for Statistical Language Modeling, Text Retrieval, Classification and Clustering. http://www.cs.cmu.edu/~mccallum/bowl.

McCallum A. and Nigam K. (1998) A Comparison of Event Models for Naïve Bayes Text Classification. In: AAAI-98 Workshop for Text Categorization.

Ranka S. and V. Singh (1998) An Efficient K-means Clustering Algorithm. In: Proc. First Workshop High Performance Data Mining.

Taghva K. and Gilbreth J. (1995) Reconizing Acronyms and their Definitions. http://www.isri.unlv.edu/publications/isripub/Taghva95-03.ps.

University of California, Irvine. Reuter 21578 Text Categorization Collection. http://kdd.ics.uci.edu//databases/reuters21578/reuters21578.html.

VanRijsbergen C. J., Robertson S. E., and Porter M. F. (1980) New models in probabilistic information retrieval. British Library, London, British Library Research and Development Report, no. 5587.

Wikipedia, The Free Encyclopedia, Data Clustering. http://en.wikipedia.org/wiki/Data_clustering.

Weisstein, Eric W. K-Means Clustering Algorithm. In: MathWorld--A Wolfram Web Resource http://mathworld.wolfram.com/K-MeansClusteringAlgorithm.html.

Extreme Programming: A Kuhnian Revolution?

Mandy Northover[1], Alan Northover[2], Stefan Gruner[1], Derrick G Kourie[1] and Andrew Boake[1]

1 University of Pretoria, Department of Computer Science mandy.northover@siemens.com
2 University of Pretoria, Department of English alan.northover@gmail.com

Abstract. This paper critically assesses the extent to which the Agile Software community's use of Thomas Kuhn's theory of revolutionary scientific change is justified. It will be argued that Kuhn's concepts of "scientific revolutions" and "paradigm shift" cannot adequately explain the change from one type of software methodology to another.

1 Introduction and Related Work

This paper aims to assess the degree to which the ideas of Thomas Kuhn, a prominent 20th century philosopher of science, can appropriately be applied to illuminate recent software engineering methodologies. In particular, "agile" software methodologies – a family of related "lightweight" methods which are fundamentally focused on being adaptable to change – will be discussed. These methodologies originated during the mid 1990s, partly as a reaction against "heavyweight" methods and partly in response to the challenges of the Internet era. Their popularity was undoubtedly established by Kent Beck's Extreme Programming (XP), which will be the specific focus of this paper.

The motivation for discussing Kuhn's ideas in this paper stems from their popularity in the recent literature of the software engineering discipline, particularly in the above-mentioned "agile" community. Several leaders of that community have referred to Kuhn's ideas, most notably the founder of XP, Kent Beck, who explicitly cites Kuhn's The Structure of Scientific Revolutions in the annotated bibliography of his own book, Extreme Programming Explained: Embrace Change.

Kuhn's revolutionary model of change in scientific methodology is used to assess the change from traditional software methodologies to "agile" methodologies. However, we shall argue that Kuhn's concepts of "scientific revolutions" and "paradigm shift" are in many ways inappropriate to explain this change.

As mentioned above, Kuhn's concepts have been applied to the software engineering discipline by several founding members of the agile project management and software development movements. However, most of those applications have been

uncritical. Before we attempt to address these shortcomings, we shall give a brief overview of some of the most significant related positions.
In (Yourdon 2001a), Yourdon references Kuhn's Structure and applies his concept of "paradigm shift" to IT organisations. In (Yourdon 2001b), he applies Kuhnian terminology specifically to XP by describing "classical" software engineering as the old paradigm and "agile", "lean" or "light" software development as the new paradigm. He also notes a presentation at the 2001 Cutter Summit conference in which Beck, himself, describes XP as a "paradigm shift".
In (Bach 2000), Bach applies the work Kuhn and several other philosophers of science to software engineering in the context of "process vs. practice". He argues that, in the 20th century, even science was a battleground for this debate when the idea of science as a rational enterprise came under fire. The resulting fallible view of science, according to Bach, is pertinent to IT today. Bach argues, perhaps controversially, that "Developing software is not much different from developing scientific theory, and developing processes for developing software is exactly like doing science". Furthermore, he argues that the trend to move from process towards practice manifests itself in methodology movements such as XP and object-oriented design.
Schwaber does not explicitly cite Kuhn in his articles but he frequently uses Kuhnian terminology. In (Schwaber 2001), he describes the seminal meeting of the Agile advocates in 2001 as a "meeting of revolutionaries". He also points out in a section of (Schwaber 2002) entitled "Viva la Revolucion", that the Agile Manifesto was a call to arms and is based on principles rather than techniques.
In (Windhotz 2002), Windholtz paraphrases several talks from the 2002 XP Agile Universe conference, most notably those of Martin Fowler and David West. Fowler argues that the Agile and XP movements are a large shift in how people think about software development and West attempts to answer, firstly, why there is an undercurrent of revolution in the Agile community and, secondly, how one should properly conduct a revolution.

2 Revolutionary Change

Kuhn's first book was entitled *The Copernican Revolution* and in it he started developing his ideas that culminated in *Structure*. In the latter book, Kuhn explicitly compares scientific and political revolutions: "Like the choice between competing political institutions, that between competing paradigms proves to be a choice between incompatible modes of community life". For Kuhn, "paradigm" means two things (Kuhn 1962):
"On the one hand, it stands for the entire constellation of beliefs, values, techniques, and so on shared by the members of a given [scientific] community. On the other, it denotes one sort of element, in that constellation, the concrete puzzle-solutions which, employed as models or examples, can replace explicit rules as a basis for the solution of the remaining puzzles of normal science."
The clearest examples of Kuhn's "paradigm shifts" come from cosmology, namely the Ptolemaic, Copernican-Newtonian, and Einsteinian revolutions.
Kuhn divides the development of science into several distinct phases, each leading ineluctably, in an endless cycle, to the next: normal science, crisis, extraordinary science, paradigm shift, new normal science.

Kuhn argues that a discipline becomes scientific only once all the practitioners adopt a single paradigm. The pre-scientific period is characterised by several competing schools of thought within a single discipline. Once a paradigm is adopted all practitioners within it become members of a scientific community, whose research is governed by the paradigm. Kuhn calls this mature science and describes the paradigm-governed activities of its scientists as normal science, which occurs uninterrupted for long periods of time. New members are induced, or initiated, into the paradigm not primarily through the explicit acquisition of the theories, rules and criteria of the paradigm but rather through studying textbook exemplars and practising in laboratories, a process of learning that Kuhn calls tacit knowledge. This is an efficient way of training new scientists in the paradigm, but it does not encourage open-minded, reflective, self-critical scientists.

Furthermore, Kuhn contends, referring to George Orwell's *Nineteen Eighty Four* that the science textbooks are re-written by members of the triumphant paradigm to suggest that all previous scientific work had evolved rationally to culminate in the present paradigm. The reason for this historical revisionism is to indoctrinate the science students and to propagate the myth of progress in science. Kuhn argues that this contradicts the historical facts.

During normal science, scientists do not attempt to refute hypotheses, theories or the paradigm, but rather engage in puzzle-solving. They resist counter-instances (anomalies) until so many have accumulated that they can no longer be ignored and the scientific community is plunged into a state of crisis. During this period of crisis, scientists practice extra-ordinary science. Kuhn argues that no scientific community will abandon the dominant paradigm unless a new one becomes available since to abandon the paradigm is to abandon science altogether.

The attainment of a new paradigm is considered revolutionary because the break with the old paradigm is final and largely irrational, since the very "rules of the game" (a term Kuhn takes from Wittgenstein) have changed. It is a complete change of worldview and is holistic rather than piecemeal. Furthermore, the new paradigm is incommensurable (cannot be compared) with the old one, so much so that scientists within different paradigms are unable to understand each other. Translation between paradigms is very difficult since the scientists belong to different "language communities" (a term also from Wittgenstein).

Kuhn doubts that science progresses as a whole towards a closer approximation of the truth, although he points out that scientists within the triumphant paradigm would consider the adoption of their paradigm as progress. For Kuhn, persuasion by argument is not decisive, but rather certain non-rational factors are required to convert people to the new paradigm, particularly faith.

Nonetheless, this emphasis on emotion towards the end of Kuhn's book betrays a shift of emphasis from the beginning where Kuhn argues that the main reason for abandoning the old paradigm was the accumulation of anomalies, namely, the paradigm's failure to solve puzzles, and that the main attraction of the new paradigm is the promise it holds to solve these puzzles. Since the new paradigm should also solve many of the old puzzles in addition to the new ones, a case can be made for rational progress in scientific knowledge as a whole especially since, for Kuhn, the main purpose of a paradigm is to solve puzzles. Indeed, the main activity of a scientist and the measure of his worth is his ability to solve puzzles.

In the following we shall assess whether Kuhn's theory adequately accounts for the change in methodology towards XP and whether the change from classical to Agile methodologies, especially to XP, can be described as a "revolution" or "paradigm shift", in the Kuhnian sense.
One can ask, as an initial critical question, whether software developers can be described as scientific researchers. It seems more obvious that programmers are engineers of sorts, since they produce something that has direct relevance to the ordinary world, whereas physical scientists are concerned with specialized research which often has no obvious bearing on everyday life. In other words, the engineer synthesizes/constructs to produce new artifacts, whereas the scientist analyzes/takes apart to acquire knowledge about an existing (natural) entity. Furthermore, scientists are not normally held accountable to the public unlike software developers who do not merely solve problems but produce software with clear applications.
Can software methodologies be seen as paradigms? Software practice can be made to fit both of Kuhn's definitions of a paradigm, at least in the broadest senses. However, what is not evident from Kuhn's definition is that he was writing specifically about scientific communities, and, even more specifically, communities of physical scientists. In its specifically scientific sense, it is doubtful that the term can be applied to software communities.
The proliferation of software methodologies since the early 1990s will surely be interpreted by the agile community in terms of Kuhn's "scientific crisis" or "extra-ordinary science". Agile methodologies are seen as the emerging paradigm that will completely replace the traditional methodologies of the old paradigm. The questioning of the fundamental principles, values and practices of software methodology that accompanies the emergence of new methodologies, compares with the similar activity of physical scientists in a state of crisis. This presupposes that software methodologies are already a "mature science" in which a single paradigm did, in fact, dominate as opposed to a "pre-scientific" state, characterised by many competing schools of thought. In fact, this does seem to be the case since, before the present crisis, the waterfall method was dominant. However, one should question the assumption that there need be a single dominant paradigm in software engineering as Kuhn argues there is in science. Cockburn's suggestion of a process per project does not seem unreasonable, although it is a thoroughly non-Kuhnian approach, since it implies that rational choices can be made between different software methodologies.
Are software methodologies really incommensurable as Kuhn alleges scientific paradigms to be? Based on numerous articles in the software literature which compare methodologies, this seems unfounded. In (Boehm 2002), for example, Boehm makes a detailed comparison between Agile and plan-driven methodologies and argues for their synthesis into a hybrid methodology. Similarly, in (Lux Group 2003), the authors claim to have successfully adapted the waterfall model by integrating key elements of approaches like "Rapid Application Development" and XP.
Was the emergence of Agile methodologies in the mid 1990s, a result of cumulative anomalies? If not, what triggered the crisis? For Kuhn, anomalies could take the form of "discoveries, or novelties of fact" on the one hand, or "inventions, or novelties of theories", on the other. Both forms of novelty, however, are not actively sought out by – and, in fact, are initially fiercely resisted by – scientific communities. Since, software methodologies do not aim to explain physical phenomena, it is

difficult to see how Kuhn's theories are applicable to them in this case. Nonetheless, a case can be made for crucial events causing a crisis in software engineering, for example, the advent of the Internet era. Object-oriented (OO) methodologies were the dominant "paradigm" at that time and most OO advocates came to recognise that developing at "Internet time" required a reconsideration of the OO process. Moreover, OO projects were often late and over budget. In order to create truly new classes of software which would deal with the challenges of the era like rapid technological change, volatile business requirements, increasing scale and complexity and shortening market time windows , the OO software process itself needed to be radically overhauled. As a result, the new Agile "paradigm" emerged and many of the OO advocates from the old paradigm became advocates of the new paradigm.

Can Agile be called a revolution if it has not already become the dominant paradigm in ten years since its initiation? Kuhn faced a similar criticism: "The period between Copernicus and Newton is often termed 'The Scientific Revolution', but the lengthy time-span involved – over 150 years – makes the process sound more like evolution than revolution" (Bullock and Trombley 1999). Whereas political revolutions seem to occur swiftly, scientific and technological revolutions can take considerable time to occur, as was evident in, for instance, the Industrial Revolution. Nonetheless, given the quick rate of change in the Information Age, perhaps the Agile revolution should have occurred already, if it is to take place at all.

3 Conclusion

Considering how questionable Kuhn's concept of "paradigm shift" is when applied to developments in software methodology, his popularity is surprising. He was not the revolutionary that the radical young American students of the late 1960s and 1970s thought he was. In fact, his critics saw him as "the official philosopher of the emerging military-industrial complex" and rather than having killed positivism, "Kuhn simply replaced the positivist search for timelessly true propositions with historically entrenched practices. Both were inherently uncritical and conformist" (Fuller 2003). As Fuller points out, part of the popularity of Kuhn is "the innocence [of his admirers] of any alternative accounts of the history of science... with which to compare Kuhn's" (Fuller 2003), such as the fallibilism of Popper or Peirce.

In fact, it can be argued that Beck shifts his emphasis in the 1st and 2nd editions of (Beck 2005) from a Kuhnian revolutionary approach to a fallibilist evolutionary one, since in the 1st edition, Beck advocates that, in order to be truly practising XP, all Agile values, principles and practices should be adopted and strictly adhered to, whereas in the 2nd edition he suggests, instead, a piecemeal approach to adopting XP. The full-scale approach towards adopting XP can be compared to Kuhn's "normal science" in which scientists uncritically accept the basic assumptions of the dominant paradigm. Also, Kuhn's insistence that paradigms are incommensurable suggests that the Agile paradigm cannot be compared to any other software paradigm. However, earlier it was shown that several authors have made such comparisons.

On the other hand, the incommensurability of paradigms implies that the choice between them would be, according to Kuhn, irrational. Presumably, Beck would disagree with this. Instead, he would surely argue that the XP practices are all rational. However, why then does he use Kuhn's irrationalist theory of scientific revolutions? Perhaps his statement in (Beck 2006) provides a clue: "Mostly in the adoption process. Kuhn helped me predict how the market would react to XP...". Since XP will require significant change and since most people fiercely resist change, the adoption of XP will not be largely due to rationality and good arguments, but will instead require an emotional conversion or leap of faith. It would seem that Beck gained this insight about the irrationality of paradigm shifts from Kuhn despite the fact that Kuhn never mentioned markets. This may be the solution: although XP's values, principles and practices are highly rational, the adoption of XP as a whole will be a matter of emotion rather than logic, as full of hype and persuasive techniques as the selling of a new product on the market. This is, however, a significant departure from the strict use of Kuhn's theory of scientific revolutions.

References

Bach, J. (2000) Process over skill. Cutter Consortium, Vol. 1, No. 8.
Beck, K., Andres, C. (2005) Extreme Programming Explained: Embrace Change. 2nd edn. Addison-Wesley, Boston.
Beck, K. (2006) Personal e-mail communication.
Boehm, B. (2002) Get Ready for Agile Methods, with Care. IEEE Software.
Fuller, S. (2003) Kuhn vs Popper: The struggle for the soul of science. Icon books, UK
In: Bullock, A., Trombley, S. (ed.) (1999) The New Fontana Dictionary of Modern Thought. 3rd edn. Harper Collins Publishers., London.
Kuhn, T.S. (1962) The Structure of Scientific Revolutions. The University of Chicago Press., Chicago.
The Lux Group. (2003) Project Lifecycles: Waterfall, Rapid Application Development, and All That. The Lux Group, Inc.
Schwaber, K. (2001) The Agile Alliance Revolution. Cutter Consortium.
Schwaber, K. (2002) XP and culture change. Cutter Consortium.
Windholtz, M. (2002) XP Universe 2002. www.objectwind.com/papers/XPUniverse2002.pdf
Yourdon, E. (2001) Paradigm Shifts. Cutter Consortium.
Yourdon, E. (2001) The XP Paradigm Shift. Cutter Consortium.

Lossy Text Compression Techniques

Venka Palaniappan[1] and Shahram Latifi[2]

1 University of Nevada Las Vegas, Electrical & Computer Eng. Dept., venka@unlv.edu
2 University of Nevada Las Vegas, Electrical & Computer Eng. Dept., latifi@unlv.edu

Abstract. Most text documents contain a large amount of redundancy. Data compression can be used to minimize this redundancy and increase transmission efficiency or save storage space. Several text compression algorithms have been introduced for lossless text compression used in critical application areas. For non-critical applications, we could use lossy text compression to improve compression efficiency. In this paper, we propose three different source models for character-based lossy text compression: Dropped Vowels (DOV), Letter Mapping (LMP), and Replacement of Characters (ROC). The working principles and transformation methods associated with these methods are presented. Compression ratios obtained are included and compared. Comparisons of performance with those of the Huffman Coding and Arithmetic Coding algorithm are also made. Finally, some ideas for further improving the performance already obtained are proposed.

1 Introduction

The text compression has played a pivotal role in archiving text information, providing noticeable saving in storage space. Lossy compression is typically applied to image data. Identification of what is irrelevant in text and eliminating this irrelevancy for text compression is the subject of this paper. The error in lossy compression techniques is defined as the difference between the original text prior to compression and the text recovered after decompression.

Lossy text compression has been studied to some degree. Semantic and generative models have been studied to give improved text compression ratio (Witten, Bell, Moffat, Manning, Smith and Thimbley 1994). Compression by taking the text as an image file and studying the pattern matching is done in (Howard 1996). Treating each of the characters individually gives more flexibility to compress the text according to the application required.

Looking into applications such as message archiving in emails, it is observed that without the exact recovery of the text, the core of the information can be retrieved. In other applications such as short hand or speedwriting, certain abbreviations can be understood almost immediately. These flexibilities give hope for lossy text compression to be effective.

The rest of the paper is organized as follows. In Section 2, the background and motivation towards developing these schemes is presented. Section 3 forms the core of the paper in which three different compression methods will be investigated. Section 4 includes the results and performance of each of the three different compressions techniques. Section 5 concludes the paper.

2 Background and Motivation

In everyday life, there are many applications that use text compression. Many studies have been carried out with the human brain and its recognition of predefined set of characters. It shows that humans have the capability of recovering the exact information being presented even if there are some character errors in their text. This yields toward a great opportunity in the world of compression.
For instance, if we drop all vowels from text, it will disrupt the readability of the entire text. But using a spell checker allows us to recover most of the text correctly, leaving a small number of words undefined. If the entire text is to be read, then the reader will be able to fill in the most suitable character into the unknown slot. This gives opportunity to explore many fields which use similar approach in the text usage. Applications such as instant messaging and emails will benefit both in the transmission time and most importantly the storage space. Online diaries or blogs are rapidly growing and these areas can benefit from lossy text compression.

3 Techniques

3.1 Letter Mapping

Excellent comprehensive dictionaries have recently become available in a machine-readable form. The Letter Mapping (LMP) method proposed here is a macro character replacement in every word in an entire text. This method takes the least frequently occurring characters and replaces them with highly occurring characters. In an English text characters like r, s, t, l, n occurs with higher probability then characters like q, z, x, v, j and also have lesser bits assigned to them when compressed using the Huffman coding. Therefore if the least occurring characters are replaced with the more frequently occurring characters, then the alphabet size in the compression scheme will be reduced, in this case by 5. Less probable characters are assigned longer bits in Huffman Coding. By changing the least occurring characters to more likely occurring characters, a higher bit rate may be achieved.
At the decoding end, the text is decompressed. The recovered text will surely contain a large number of errors due to misspelled words. By running the text through a simple spell checker, it will rectify most of the errors in the text. It will be easy to decide on the choice of words because words in error are actually words with the characters (q, z, x, v, j) in them. Therefore, the spell checker only needs to look for the words with these characters in them. Clearly this step does not guarantee full recovery of the text. Some recovered words with changed characters will not be picked up by the spell checker. This is because the newly reconstructed words will have a proper spelling e.g ('save' after LMP will be 'sane').
Different text documents are selected to carry out this compression to investigate which application is most suitable for the lossy text compression with an acceptable

error rate. The assignments of characters for replacement are also varied to check the most efficient text recovery.

3.2 Dropped Vowels

In English text, vowels occur very frequently throughout the text. If the vowels were dropped from the text, a significant increase in the compression is achieved due to their frequency of occurrence. But with all the vowels dropped, the problem of recovering the word again will be much harder. With this concern, the vowels are all changed to blanks "□" or to alphabet "e" ($a \rightarrow e, o \rightarrow e, i \rightarrow e, u \rightarrow e$). The latter of the two will give a better error recovery, but both schemes give the same compression rate and bit rate. This method will reduce bit rate because instead of using 26 characters, only 22 are used after encoding. At the decoding end, the algorithm adopted is similar to the LMP but differ only in the second step. It will substitute all the vowels with the character "e".

3.3 Replacement of Characters

Sacrificing accuracy for compression is not suitable for practical use. But to sacrifice accuracy to an extent where it is still readable to the user may be of use in certain type of applications. The human mind will be able to make up the words present in the text even though some characters are missing but it denotes the same sound as the original word. Replacement of Characters (ROC) is done based on a commonly used method in writing called shorthand or brief-hand. Using this technique, a combination of several characters can be represented as one character, thus reducing the number of characters in a text. In ROC capital letters / upper case letters are replaced with their equivalent lower case letters and thereby removing 26 characters from the alphabet size. This will give an improvement to the bit rate as well.
The recovered text will have many errors present on it but the text is acceptable if it is readable. This type of lossy compression is more suitable for applications such as Text Messaging, Online Journal, Blogs, and Email Archiving. For this type of applications the text itself is not as important as reading the information in it. Therefore, this technique uses less space to store the same amount of information.

4 Analysis

4.1 Compression Analysis

The performance of LMP, DOV and ROC are compared here. The trade-off between the compression factor and accuracy is analyzed. The compression was done using Huffman Coding. The test sets were taken from various text data. Test sets 1 & 3 are from journals. Test sets 4 & 5 are from history articles. Test sets 6 & 7 are taken from online journals or blogs. Test sets 8 & 10 are from technical paper. Test set 9 is from children's story book.
From the testing carried out using different data sets, it is found that the improvement ratio varies for different type of application. Table 1 shows the percentage of improvement for different lossy compression techniques with respect to lossless compression using both Huffman Coding.

TS	LMP+Huff	DOV+Huff	ROC+Huff
1	(1.53%)	0.49%	(3.81%)
2	0.72%	5.21%	8.14%
3	0.10%	7.85%	8.73%
4	2.53%	11.04%	14.14%
5	5.28%	14.24%	12.15%
6	6.53%	11.41%	14.19%
7	5.65%	13.05%	14.98%
8	1.93%	5.42%	4.31%
9	6.53%	11.09%	13.32%
10	0.20%	4.81%	7.52%

Table 1. Compression Improvement Using LMP, DOV, ROC Encoding + Huffman Coding

The file sizes are presented in bytes and the improvement in percentages. The original refers to the file after compression without using any transformation. The LMP+Huff, DOV+Huff and ROC + Huff each corresponds to the lossy transformation used and the compression techniques used.

4.2 Theoretical Analysis

Initially an unaltered text will have an alphabet size with all the characters present $A = \{a, b, c..., z\}$. Each of these characters will have its individual probability of occurrence. When Huffman Coding is applied, each of these characters in the alphabet will be assigned a unique code. The performance of the code is measured using the entropy (H) and the average length (l). The entropy is,

$$H = -\sum P(A_i) \log_2 P(A_i) \quad (1)$$

= 4.1758 bits/symbol

The bit rate is,

$$l = \sum (P(A_i) \times l_i) \quad (2)$$

= 4.1847 bits/symbol

A measure of efficiency of this code is its redundancy,

$$R = l - H \quad (3)$$

= 4.1847 – 4.1758 =0.0090 bits/ symbol

When Letter Mapping (LMP) is applied to the text, the alphabet size reduces by the number of characters being mapped $A = \{a, b, c...,\}$ & $A \neq \{z, q, x, v, j\}$. The performance of the code is better than the initial code.

The entropy is, H_{LMP} = 4.1022 bits/symbol. The bit rate is, l_{LMP} = 4.1373 bits/symbol. The redundancy is, R_{LMP} = 4.1022 – 4.1373 =0.0351 bits/ symbol

The Dropped of Vowel (DOV) technique yields a higher performance because characters with higher probability is removed and replace with "e", which is the highest probability of occurrence. Therefore the number of bits required to code the alphabet will be far less. The entropy is, H_{DOV} = 3.3418 bits/symbol. The bit rate is,

l_{DOV} = 3.4040 bits/symbol. The redundancy is, R_{DOV} = 3.3418-3.40490 =0.0622 bits/ symbol

The improvement in compression is proportional to the occurrence of error in the text after decompression. As shown in Figure 1, the numbers of errors increases for higher compression ratio.

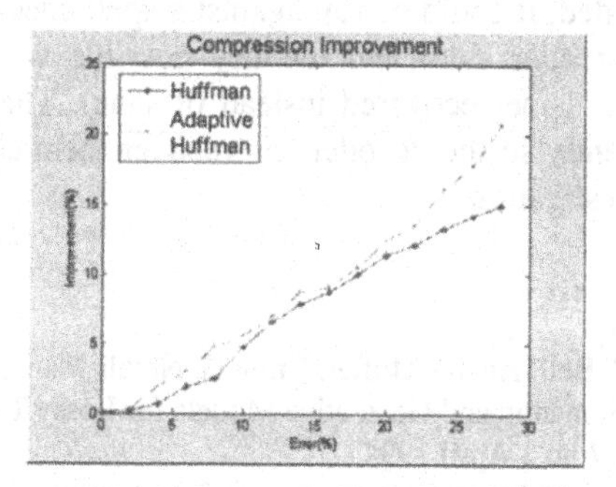

Fig. 1. Compression Improvement

To achieve higher compression ratio, the accuracy of the recovered text is sacrificed.

5 Discussion and Conclusion

Improvement can be made to reduce the number of errors by introducing a spell checker in the encoder's end to mimic the decoder's spell checker. This allows reduction of misinterpretation errors. By having the spell checker in the encoder's end, it will check for difference between the words which have undergone the transformation with respect to the original text. For instance, taking the word "save" and applying letter mapping to it will result in a transformed word "sane". This word will go undetected in the decoder's spell checker and result in an error. Instead, if we have a spell checker at the encoder's end which will detect all the words which have been transformed with respect to the original text and sending this information to the decoder's spell checker, misinterpretation errors would be reduced. Now taking the same example, the word "sane" will be detected by the encoder's spell checker and this information will be passed on to the decoder's spell checker. Decoder's spell checker will look for other spelling options for the word "sane" which will result in "save" and correctly recovering the word.

The techniques which have been described result in a modest but worthwhile improvement. From observation, it can be concluded that the compression ratio for lossy text compression is not as large as expected in contrast to the lossy image compression. This is due to fact that in a text there are many different permutation and combination occurrence of the characters. Therefore there are many limitations to finding an algorithm which can give a high compression ratio as the lossy image compression does. Another limitation encountered is the application for lossy text compression. From the result obtained, the data set which gives the highest improvement factor are those from standard text, which does not have many technical or specific terms in them. These kinds of text are found in normal reading material such as story books and newspapers. But the field which could benefit mostly from lossy text compression would be archiving of online journal, blogs, text messaging and emails. The result shows that Replacement of Characters yields the best compression ratio and this scheme does not suffer from decoding errors.

Future work includes combining two of the techniques into one hybrid technique and analyzing the new scheme. Another possible improvement to the schemes introduced here is to have the encoder mimic the action of the decoder. More specifically, once the information is encoded, it could be run against a spell checkers (this is what will also be done at the decoder side) and identify possible words which cannot be correctly recovered (ex. same recovered instead of sane). These special words can thus be encoded differently so the decoder can recover them correctly. This issue is currently under our investigation.

References

Ian H. Witten, Timothy C. Bell, Alistar Moffat, Craig G. Nevill-Manning, Tony C.Smith and Harold Thimbleby, "Semantic and Generative Models for Lossy Text Compression", The Computer Journal, v.37 no.2 April 1994

Paul G. Howard, "Lossless and Lossy Compression of Text Images by Soft Pattern Matching", IEEE Transaction, pg 210-219, 1996

Micheail Pechura, "File Archiving Techniques Using Data Compression", Communication of the ACM, v.25, n.9, 1982

Yair Kaufman and Shmuel T. Klein, "Semiloss Text Compression", Proceedings of the Prague Stringology Conferenc, 2004

Self-Customization of E-Knowledge Using Color Semantics

Azita Bahrami, Ph.D.

Armstrong Atlantic State University, Department of Information Technology, 11935 Abercorn Street, Savannah, GA 31419, bahramaz@mail.armstrong.edu

Abstract. It is a common practice for readers to highlight portions, *clumps*, of a printed document deemed important for later use. In this research effort, the benefits of highlighting process are extended to e-Documents. The analysis, design and implementation of a software system, *Prospector,* is presented, which is able to (1) provide for highlighting of clumps in an e-Document using different colors, (2) enforce color semantics, (3) deliver customized veins of clumps in an e-Document, (4) aggregate the same-color clumps, (5) support dynamic review of the highlighted clumps, and (6) provide for editing of the clumps.

1 Introduction

When reading a hardcopy document for learning purposes, the reader is likely to use a highlighting pen to highlight portions of the document that are important to him/her. The reason is that the highlighted portions (paragraphs, sentences, phrases, or words,) are the jest of the printed document suitable for revisiting later for mastery learning. Often, readers use more than one color for highlighting to (1) separate various concepts within the same document and (2) give different degrees of importance to the contents of a document as a tool for memorization purposes, among other reasons. For instance, the "advantages" of a scientific approach on page 5 of a document can be highlighted pastel yellow. The mention of another vital but implicit "advantage" on page 20 can also be highlighted pastel yellow for later addition to the list of advantages. The "disadvantages" may be highlighted pastel blue anywhere they appear in the text. At the conclusion of reading the document, there may be several sets of highlighted texts in different colors signifying different topics within the same document. These sets may be aggregated later through handwriting or typing.
In this paper, each highlighted section of the document is considered to be a *clump of knowledge* or *clump,* and the clumps collectively along with their relationships are considered to be the *vein of clumps* or *vein*. The vein of a document may be different from one reader to the next.

Though using different-color highlighters for highlighting sections of a hard copy is relatively fast and easy, creating the document's vein and aggregating the same-color clumps by handwriting or typing is quite time consuming.

This research effort proposes extending the benefits of highlighting process to e-Documents, such as on-line/off-line textbooks, magazines, newspapers, reports, articles, web pages, etc. To accomplish this, a software system, Prospector, is analyzed, designed, and implanted, which (1) provides for highlighting of clumps in an e-Document using different colors, (2) enforces *color semantics*, (3) delivers a customized vein of an e-Document, (4) aggregates clumps of the same colors, (5) supports dynamic review of clumps, and (6) provides for editing of the clumps.

The organization for the rest of the paper is as follow. The Related Work is discussed in section 2, Methodology is presented in section 3, Implementation of the system is covered in section 4, and Summary and Future Research are presented in section 5.

2 Related Works

The body of existing work on highlighting of e-Documents can be grouped into two major categories of *system-driven* and *user-driven*. In the first category, a user asks for a specific token or a string of tokens in an e-Document causing the system to search the e-Document and highlight all the instances of the asked token(s). The highlighting process may have several variations. For example, the highlighting of tokens may happen one at a time or all at once. In either case, more than one color may be used, one per token. Debuggers, search engines, and word processing systems are examples of works in this category (Marcu and Abe 1995; Hull and Lee 2000; Oracle Corporation 2001; Brown, Burbano, Minski and Cruz 2002; Pace 2003; Preston, Preston, and Ferrett 2003; Riggott and Suda 2004).

In the second category, the highlighting process serves totally a different purpose. The highlighting is done by the reader using a mouse or a keyboard to mark portions of an e-Document that are important to him/her. One of the best examples in this category is the TextMarker! (Mozilla Company 2006), which enables a reader to highlight important portions of a web page and save them into a file, if so desired.

Prospector falls into the second category. The major differences between TextMarker! and Prospector are that the former is not capable of enforcing color semantics in the highlighting process, finding document's vein, providing for dynamic review of clumps, or supporting a comprehensive editing capability.

3 Methodology

In this section, the concept of *Color Semantics* is presented followed by *Highlighting Modes, Vein Representation*, and *Dynamic Review of the Clumps*.

3.1 Color Semantics

The colors are semantically divided into two finite sets of *background* and *foreground* colors. Each topic of interest to the reader within an e-Document is highlighted with a different background color indicating that one topic is different from the next. Each sub-section of interest within a given topic (highlighted by a background color) may be highlighted by a foreground color indicating that the sub-sections have different levels of importance.

The combination of the following rules composes the concept of Color Semantics.

1. Each background and foreground color has a unique identification number.
2. Each background and foreground color has an order number.
3. The order numbers of all the background colors are zero.
4. The order numbers of foreground colors are different from one color to the next and start from one.
5. An order number signifies the importance of a highlighted section. The higher the order number, the more important is the highlighted section.
6. The clumps with the same background color are related.
7. There is a hierarchy of colors; the lowest being the clump's color, which is the background color in pastel followed by sub-clumps' colors. The innermost sub-clump color, therefore, has the highest value (the most important concept within an already important segment of an e-Document.)

Example: The *pastel colors* are used for background, *colors* are used for foreground, and the order numbers of the three foreground colors of green, red and blue are respectively 3, 2, and 1. A particular concept whose various aspects are scattered throughout the text may be considered as Topic A. This topic may be highlighted with pastel yellow (a paragraph, for example) and is considered as a clump. A sub-clump, a *nugget*, within the highlighted yellow area (a sentence, for example) that represents more important value than the paragraph as a whole, is highlighted with a non-pastel blue. Within the blue nugget a more important nugget (a phrase, for example) is highlighted red. Finally, within the red nugget, the most important segment (a word, for example) is highlighted with non-pastel green.

3.2 Highlighting Modes

In an e-Document, D, a clump may be *simple* or *composite*. If simple, the entire clump is highlighted in only one background color with no parts of it layered by other colors. If, however, the clump is composite, it is first highlighted by a background pastel color and then the more important concept within the clump is highlighted by a color based on the color semantics prescription. The reverse hierarchy of the highlighted sections is as follow. Pastel color (a paragraph, for example), blue (a sentence, for example), red (a phrase, for example), and green (a word, for example). Therefore, the color of the innermost sub-clump, nugget, has the highest value.

Let $BC_1 \ldots BC_n$ be a set of finite background colors used for various topics of interest within an e-Document. Also, let $FC_1 \ldots FC_m$ be a set of finite foreground colors used for various importance levels of sub-topics (nuggets) of an e-Document. The enforcement of color semantics is manifested in one of the following highlighting modes. A user may select a highlighting mode that suits his or her needs.

Just-A-Color (JAC): Colors are used randomly for both background and foreground. Thus, their order numbers are totally ignored. The selected colors are considered as background colors, whether pastel or non-pastel. This is very similar to highlighting with random colors when marking segments of a hard copy document with no significance attached to the chosen color.

Background Identification Color (BIC): Different pastel colors are used to indicate different topics. The same topic scattered in various locations in an e-Document is

highlighted with the same color. The identification number of the background colors is used for subsequent aggregation.
Foreground Hierarchy Color (FHC): Foreground colors are used based on their order number to express the *importance level* hierarchy among the nuggets of a clump. In this paper, the assumption is that FC_1 and FC_m, respectively, have the lowest and the highest order numbers.
BIC-and-FHC (BIC-FHC): The combination of BIC and FHC provides a thorough control over all highlighted portions, making the aggregated output most meaningful as the result of providing a concise collection of sets of related concepts along with their degrees of importance.

3.4 Vein Representation

Each new layer of color within a clump is logically a clump by itself, though it is referred to as a nugget. Thus, each clump has the following properties:
Importance Level is an integer number, whose value is equal to the value of the color order number of the clump. The higher the number, the more important is the clump. ***Depth Level*** is an integer number whose value is initially equal to one (initial depth level) for every clump. The final depth level for each clump is equal to the sum of its initial depth level and the depth level of its immediate embedding clump. As a result, the final depth level of the *outermost* embedding clumps remains one whereas the depth level of each embedded clump is recalculated.
The depth levels of clumps in an e-Document are used for construction of the document's *vein.* Such a vein has a tree structure and represents the relationships among all the clumps and the nuggets of an e-Document. The Name of this clump tree is the name of the e-Document. All of the clumps with the depth level of 1 make up the first level of the tree. All of the nuggets of a clump with the same depth level are siblings and they are children of the embedding clump.
The vein of the clumps highlighted in the BIC and BIC-FHC modes do not have a simple tree structure. Instead, their vein structures are clump trees in which the nodes of the same background colors are linked together to fully support the aggregation process. The clump trees for the e-Document, D, are illustrated in Figures 1.a, 1.b, and 1.c, 1.d, and 1.e where JAC, BIC, FHC and BIC-FHC highlighting modes are respectively used.

3.5 Dynamic review of the clumps

The clump tree of an e-Document may be *static* or *dynamic.* A static clump tree is simply a graphical representation of an e-Document's vein in which each node represents a clump. In addition to that ability, a dynamic clump tree is able to reveal its contents up on a click by the user. Such a feature enables the reader to see the actual contents of a clump within the clump tree and edit the contents of each node. The major benefit of having this editing feature in the dynamic clump trees is that the user can incorporate his/her annotations to the nodes and, thereby, fully customize them.

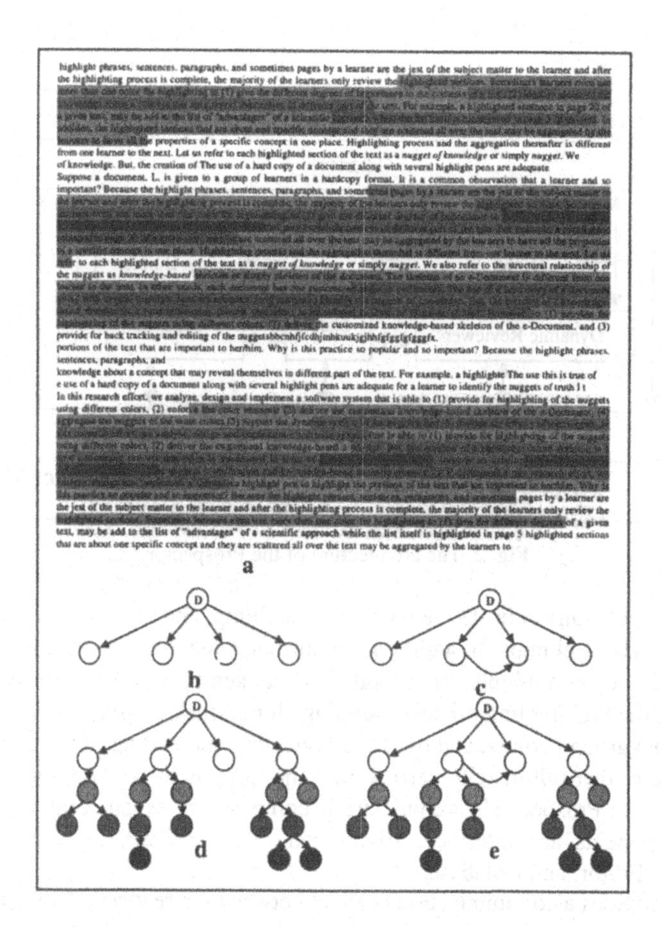

Fig. 1. An e-Document and four of its clump trees: (a) An e-Document, (b) Clump tree for JAC, (c) Clump tree for BIC (d) Clump tree for FHC, and (e) Clump tree for BIC-FHC highlighting modes.

4 Implementation

An e-Document can be considered as a string of characters. In addition, each figure or table (excluding their captions) can be considered as a character within an e-Document. Such considerations translate into assigning an order number to each character. The character order number is an integer starting from one. As a result, each clump (and nugget) is represented by four attributes of *Color (C), Importance level, (L), Starting position (S)* and *Ending position (E)*.

The attribute C represents the highlighting color chosen by the reader for the clump. The attribute L carries the importance level associated with the clump. The attributes S and E respectively represent the order numbers of the first and the last characters of the clump. A database is utilized for depositing the attributes of each clump.

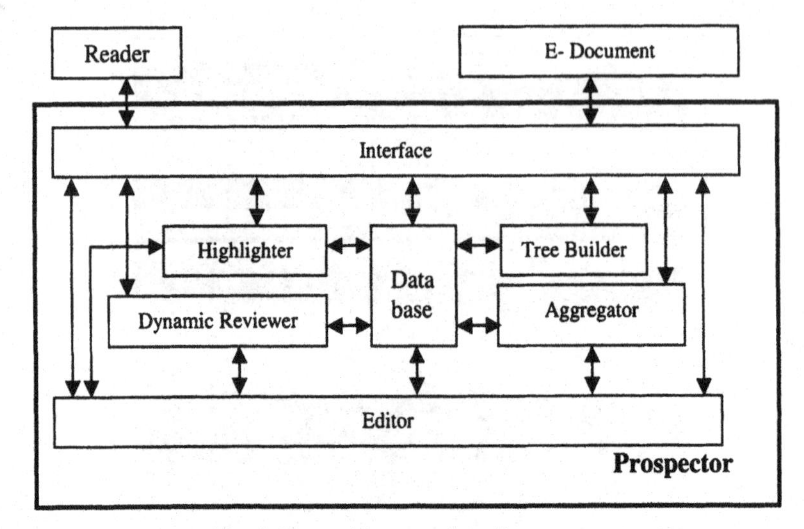

Fig. 2. The architecture of the Prospector.

The S and E positions indicate the location of a clump, and they are used to calculate the depth levels of clumps. In addition, all attributes together are able to provide for all the needs of every highlighting mode. Consequently, the database mentioned is vital during the highlighting process, building clump trees, aggregating the clumps of the same background colors, and dynamic reviewing and editing of the clumps.
The software that ultimately carries out the proposed methodology is named Prospector. The Prospector's architecture is composed of seven components, Fig. 2. These components are Interface, Highlighter, Tree Builder, Dynamic Reviewer, Aggregator, Editor, and Database.
Interface provides a communication channel between the reader and Prospector.
Highlighter supports the selection of highlighting colors and collects the clump attributes.
Tree Builder constructs clump trees of an e-Document for a reader using the reader's selected clumps and highlighting mode.
Dynamic Reviewer enables the user to click on each node of the tree (delivered by the Tree Builder), review, and edit the actual clump for which the node stands.
Aggregator aggregates all of the clumps with the same background color into an aggregated file.
Editor assists the reader in editing of clumps during: (1) highlighting process: deleting and/or saving highlighted clumps, (2) dynamic review process: editing actual contents of each node of a dynamic clump tree, (3) aggregation process: editing the contents of an aggregated file, and (4) dynamic review: permitting the reader to add annotations for each node.
Database is a depository of the clumps' attributes.

5 Summary and Future Research

It is a common practice for readers to highlight portions, *clumps*, of a printed document deemed important for later use. A software system, Prospector, is analyzed, designed, and implemented to extend the benefits of highlighting process to e-Documents. Four different highlighting modes are introduced that each one enforces color semantics differently: (1) in JAC the Colors are used randomly for both background and foreground and their order numbers are totally ignored, (2) in BIC in the highlighted clumps of the same color are grouped, (3) in FHC the colors are prioritized by assigning an order number to each color, and (4) in BIC-FHC the combination of 2 and 3 are used. Prospector supports all four highlighting modes and is able to deliver relationships among clumps of an e-Document for each mode in the form of a clump tree. Each document has one customized clump tree (e-knowledge) per reader.

In addition, Prospector provides for aggregation of clumps with the same background color clumps, supports the dynamic review of clumps within a clump tree, and provides for editing of the clumps at different stages of highlighting, aggregating, and dynamic reviewing.

As future research, the use of the clump trees for determining the logical structure of an e-Document is in progress.

Acknowledgement

I would like to extend my special thanks to my student, Mr. Casey Zebrowski, for his successful coding of the Prospector.

References

Brown, D. C., Burbano, E., Minski, J., and Cruz, I. F. (2002) An evaluation of the effects of web page color and layout adaptations. IEEE Multimedia, Vol. 9, No. 1, pp. 86-89.

Hull, J. J. and Lee D. (2000) Simultaneous Highlighting of Paper and Electronic Documents. 15th International Conference on Pattern Recognition, Vol. 4, p. 4401.

Marcu, G. and Abe S., (1995) Blue-Print Document Analysis for Color Classification, Proc. of International Conference on Image Analysis and Processing, San Remo, Italy, pp.569-579.

Mozilla Company (2006) TextMarker!. (Beta release), Version 0.3.2, https:// addons. mozilla.org/firefox/559.

Oracle Corporation (2001) *Oracle Text: An Oracle Technical White Paper*, Oracle Corporation, pp. 16-17.

Pace, J. (2003) Text::Highlight. Beta release, Version 1.0.2, http://search.cpan.org/~icrf /text-highlight-0.04/lib/text/highlight.pm

Preston, J., Preston, S., and Ferrett, R.F.(2003) *Learn Office XP*. Prentice Hall, New Jersey, Vol. 1, pp.143-145.

Riggott, M. and Suda, B. (2004) Enhance Usability by Highlighting Search Terms. Scripting, Issue 186, http://alistapart.com/topics/code/scripting/

Section 6

Industry Day 2007

A Transactions Framework for Effective Enterprise Knowledge Management

Simon Polovina, Richard Hill

Communication & Computing Research Centre
Faculty of Arts, Computing, Engineering & Sciences
Sheffield Hallam University, UK S1 1WB
{s.polovina,r.hill}@shu.ac.uk

Abstract. An enterprise's knowledge of itself and its environment can be greatly enhanced if it can access and integrate its divergently encoded data, information and knowledge bases. A framework is proposed that is based around transactions. Transactions recognise the optimal relationships that occur through a mutually beneficial exchange in resources. Using conceptual graphs as the lucid illustration vehicle, a simple but illuminating case study is provided. It shows how transactions can transcend the myriad quantitative and qualitative aspects of an enterprise. The transaction framework thereby provides enterprises with a useful organisational memory, thereby enabling more effective enterprise knowledge management.

1 Introduction

For effective knowledge management to be realised in computer-based applications that enterprises can use in practice, there has to be some common basis by which that enterprise's knowledge of itself and its environment can be accessed and integrated across all its divergently encoded data, information and knowledge bases.
We propose a framework that is based around transactions. Transactions recognise the consequential benefits and costs that occur through the exchange of resources, for example a willingness to give up cash or take on loan debt for the overall perceived benefit of owning a motor car.

2 A Practical Demonstration: P-H University

To illustrate our argument we provide a simple but illuminating practical demonstration of how transactions transcend the myriad quantitative and qualitative aspects of an enterprise, its transactions with other enterprises and the environment. Using Sowa's conceptual graphs as the lucid illustration vehicle, we work through a short case study involving a fictional University called 'P-H University' [1,2,3].

P-H University has a student population of 15,000 and an annual turnover of £15m (15,000,000 British pounds). It specialises in technological subjects, with centres of excellences in certain areas. Due to uncertainties in government policy, students' preferences for non-technological courses, increased staff and equipment costs, and an increasingly competitive higher educational market it has a difficult year and is expected to remain so for the next two years. Indeed this year the university will make a loss of £1m.

The university's staff are concerned about keeping their jobs, not helped by the equivocal statements given by management who are in turn pressed by the financial statements that paint a grim picture. Consultants to the university have advised that the situation is cyclical, as there is an emerging trend by industry that wish to recruit technological graduates, as well as a significant increase in interest by schoolchildren in technology after a number of successful initiatives by government and industry. The university's management are nonetheless concerned that the university will not survive the current cycle, which they view as uncertain anyway, and has suspended all staff development and is seriously considering applying the same to the research budget for emerging researchers who do not yet generate income. The university is beginning to lose key staff who simply choose to leave, and risks losing credibility amongst its community and its profile in higher education overall. But by saving these costs a net surplus of £1m can instead be made, further increased by the salaries saved (allowing for pay-offs such as redundancy or other associated costs) of those staff leaving.

Many of P-H's staff are research active. This means they pride themselves on the quality of their research. 20% of the staff generate 80% of the research output. They bring in a substantial amount of research income that contributes £7m to the bottom line. A further 40% are emerging researchers contributing the remaining 20% of the research output but little that is income generating presently. It is this group that are most affected by the proposed research budget cut and although most of these staff are resigned to this fate, it will have a significantly adverse impact on their motivation. This will have an effect on P-H that presently cannot be calculated but is worryingly adverse. The other 40% of staff are interested in teaching only and do not contribute to research, but are already de-motivated by the loss of staff development. As many of them aspire to be research active, the loss of psychical enjoyment offered by this career path, like those already engaged in research is incalculable.

The university's Director of Research and representatives of the research staff meet to decide what the best course of action should be. They determine the situation as captured by the view highlighted in the conceptual graph of Figure 1. Conceptual graphs are used as they capture the concepts by which humans can articulate their thinking whilst allowing a direct mapping into computer structures for processing, such as first order logic. The basis of this graph is the conceptual graph Figure 2, being Polovina's Economic Accounting model that in turn is based on Geerts & McCarthy's respected Event Accounting model [4,5,6]. We see that this model captures the full essence of a transaction.

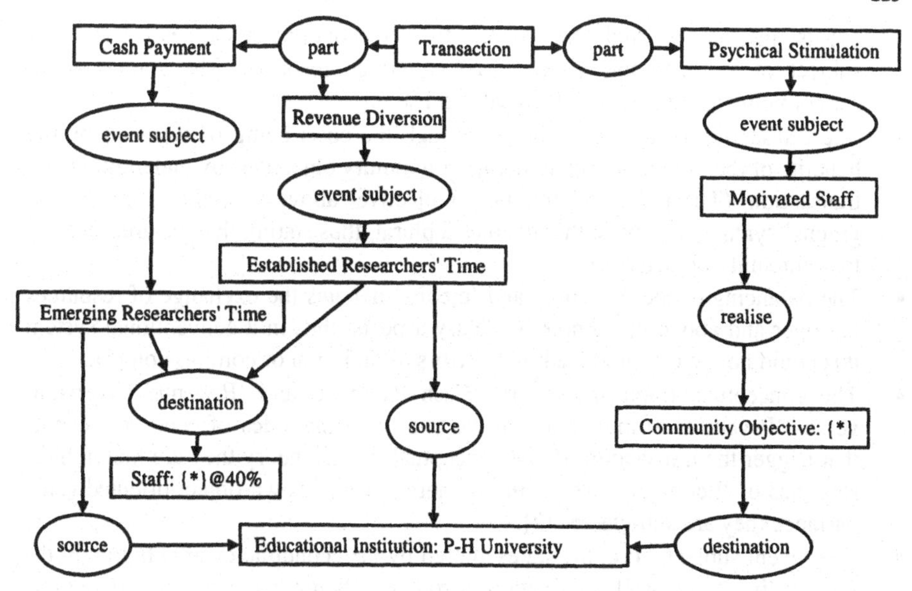

Figure 1. View of P-H University scenario

Note that Figure 1 reveals not only the transaction's monetary dimensions but the qualitative dimensions, too. These added dimensions have been captured by this particular conceptual graph thus can now be added into the university's computerised organisational memory, due to the functionality of conceptual graphs as described earlier.

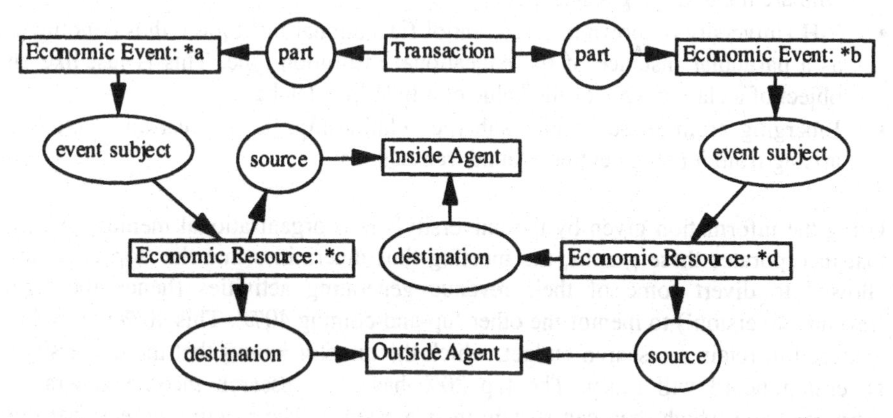

Figure 2. Polovina's economic accounting model, after Geerts et al.'s Event Accounting.

Without going into the full richness of the conceptual graphs theory, which space precludes us from doing so here, relating Figure 2 to Figure 1 we observe the following:

- Conceptual graphs are made up of concepts (the boxes) and relations (the ovals) linked by directed arrows. They may be read as "a/the *(oval)* of a *[box at the tail of the arrow]* is a *[box at the tail of the arrow]*". Thus a part of a transaction is an economic event. (The '**a*' and '**b*' denote they are separate economic events, just as '*c*' and '*d*' denote distinct economic resources.) The event subject of an economic event is an economic resource. The source of an economic resource is an inside agent. And so on.

- The transaction reveals its validity through the costs being outweighed by the benefits of the university achieving its community objectives by undertaking this transaction. (The *{*}* in 'Community Objective' above is simply a conceptual graphs' syntactical device that denotes a plural, thus stating that we are referring to community objective*s*.)
- The balancing of these 'debits' and 'credits' denotes the exchange of resources but over and above the simple monetary aspects, thus in a conventional system this could not be captured leading to errors of omission or commission [5].
- The conceptual graph shows that 'Cash Payment' and 'Revenue Diversion' versus 'Psychical Stimulation' are the two opposing sides of economic events that trigger the transaction. In the conceptual graphs theory they are hierarchical subtypes of 'Economic Event'; in the more widely known object-oriented (OO) parlance they are sub-classes [7].
- The 'event subjects' (i.e. the states altered by the economic events) highlight the salient time and staff motivation resources (being subtypes of 'Economic Resource').
- The sources and destinations (i.e. providers and recipients) of the resources are the agents in the transaction. The subtype of 'inside agent' in this scenario being the educational institution, with the 'outside agent' being the staff involved as its corresponding subtype.
- The *{*}@40%* once again describes a plural of staff, but in particular the 40% who are the emerging researchers.
- P-H University is shown as a referent of Educational Institution, thus denoting it as a particular instance of the educational institution type. This is just like an object of a class in OO, or the value of a field in a database.
- Emerging from these simple subtype relationships is a consistent ontology arising from it being centred around transactions.

Using the information given by the university's new organisational memory system that incorporates these graphs, the meeting thus determines that the 'top 20%' are allowed to divert some of their revenue generating activities (hence the term 'revenue diversion') to mentor the other 'up-and-coming 40%'. This 40% in turn has managed to retain a research budget, which the director knows that the university's governing board will ratify. The top 20% have the research income generating activities from which they can sustain their existence. The meeting agrees that this provides the most conducive environment to motivate the staff (who are thus more appreciative of the difficult environment), sustain the university in the present difficult climate and grow it in the future according to its community objectives. The university will show a net loss of £0.5m (500,000 British pounds) but this is now considered the optimal worthwhile investment for achieving its community objectives whilst retaining its sound financial management. Whilst this example contains quantitative measures for simplicity of this illustration, the real power of this approach is demonstrated when qualitative measures are the norm, as highlighted by the Transaction Agent Modelling (TrAM) methodology [8].

From the P-H University demonstration we can begin to appreciate that the transaction framework provides enterprises with an underlying basis by which they can develop and integrate their divergent systems into a truly interoperable organisational memory that in turn enables effective enterprise knowledge management.

3 Concluding Remarks

Using Sowa's conceptual graphs as the lucid illustration vehicle, we have revealed how transactions transcend the myriad quantitative and qualitative aspects of an enterprise, its transactions with other enterprises and the environment. The transaction framework thereby provides enterprises with an underlying basis by which they can interrogate their divergent systems, build an organisational memory and thereby enable effective enterprise knowledge management.

References

Gerbé, O., Keller, R. K., Mineau, G. W.: Conceptual Graphs for Representing Business Processes in Corporate Memories. In M. Mugnier, & M. Chein (Eds.), Conceptual Structures: Theory, Tools and Applications: 6th International Conference on Conceptual Structures, ICCS'98, Montpellier, France, August 1998 (proceedings). Heidelberg: Springer-Verlag (1998) 401-415

Sowa, J. F.: Knowledge Representation: Logical, Philosophical and Computational Foundations. Brooks-Cole (2000)

Sowa, J. F.: Conceptual Structures: Information Processing in Mind and Machine. Addison-Wesley (1984)

Geerts, G., McCarthy, W. E.: Database Accounting Systems. In B. Williams, & B. J. Sproul (Eds.), Information Technology Perspectives in Accounting: An Integrated Approach. Chapman and Hall Publishers (1991) 159-183

Polovina, S.: Bridging Accounting and Business Strategic Planning Using Conceptual Graphs. In H. D. Pfeiffer, & T. E. Nagle (Eds.), Conceptual Structures: Theory and Implementation. Berlin: Springer-Verlag (1993) 312-321

Polovina, S.: The Suitability of Conceptual Graphs in Strategic Management Accountancy. PhD thesis, Loughborough University (1993)

Shan, Y., Earle, R. H.: Enterprise Computing with Objects: From Client/Server Environments to the Internet. Addison Wesley (1998)

Hill, R. (2007...

Data Analyzer Prototype Using a K Data Structure Based on the Phaneron of C. S. Peirce

Jane Campbell, Mazzagatti
Unisys Corporation, Blue Bell, PA 19424
jane.mazzagatti@unisys.com

Abstract. Analyzing large data sets has always been a challenge. This paper describes a prototype of a data analyzer which uses a new data structure technology to organize a data set for analysis. This data structure, called a K or Knowledge Store, is not a copy of the data set but a recording of the relationships between the elements of the data set. The K data structure resides in memory and is capable of supporting all analytic functions.

1 Introduction

A rudimentary triadic structure, based on the Peirce phaneron, that creates a recording of basic data and relational information, has been described and implemented [KS][PT]. An API has been developed, to this software, that facilitates the creating and querying of this triadic structure called a Knowledge Store or K structure, for field/record data. Also, a GUI interface that uses the KStore API has been prototyped, for data analysis. This system is capable of returning the number of occurrences of any SIGN within the context or constraints of any set of SIGNs and therefore will support any type of analysis, of the data recorded in the K.
This prototype is not a complete product. It was implemented as a demo of the potential of the K data structure. The prototype includes a simple classification analytic, a Naïve Bayes classification analytic, a graphing tool, as well as an aggregation function that can create re-aggregated forms of the original data.
The KStore data analysis system is very efficient for building the K structure and for querying, large field/record data sets. It can process static or streaming data and does not require any external structure such as cubes or indexes.

1.1 KStore Data Analysis System Architechture

Fig. 1 shows the components of the KStore data analysis system prototype. The complete system can be installed on a PC or a server, or the server layer can be installed on a server and the application layer on client PCs.
The K data structure can be accessed by multiple instances of the K Engine and therefore can process input and queries from multiple sources simultaneously.
The K Engine processes a stream of data that has been formatted by the Learn Engine or an API Utility. As each data particle is processed the K Engine traverses

the K data structure and returns a K location to the calling routine. The calling routines then use the K locations to extract information from the K. K locations cannot be accessed by softwares in the application layer.

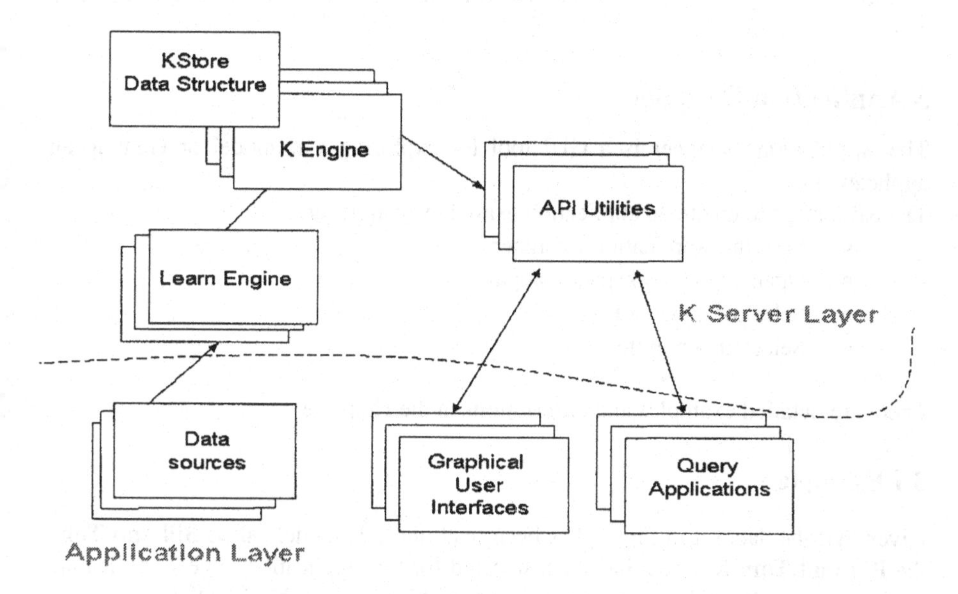

Fig.1

2 Data Sources

The K resides in the computer memory. The amount of data that can be loaded is limited only by the particular computer system resources.
Because a K models the data presented to it, instantiated Ks are called K Models. Multiple K Models can be maintained and are referenced by the K Model name.

2.1 Data Loader

The Data Loader is an application that can be accessed through the Application Designer, Administration menu and allows for the specification of .csv files to be read into the K Model.
Multiple data sources can be specified and separate threads will started for each data source.
Two data source applications have been prototyped: the Data Loader and the Data Simulator.

2.2 Data Simulator

The Data Simulator is an application that can be accessed through the Application Designer, Administration menu. It allows for the definition of a streaming data set, which will be simulated one record at a time and read into a K Model. Briefly, the

Data Simulator requires the specification of the field names, the values for each field and the probability of each field value.
Multiple Data Simulators with different parameters can be run simultaneously to stream data to a particular K. This allows for creation of very complex dynamic data sets.

3 Application Designer

The Application Designer is a GUI tool for exploring a K model or creating an application.
The basic steps to create an application using the designer are:

- Start the Application Designer
- Open a new workspace (tabpage)
- Select or create a K model
- Select an Analytic

There are detailed examples and descriptions in the Help file.

3.1 Example

Given that the data set in Fig. 2, has been read into a K model called **Bill and Tom,** the **Bill and Tom** K model has been selected for this application, a new application workspace (TabPage) has been created and the **Single Variable Prediction** analytic has been selected, the Application Designer screen will be as the screen in Fig. 3.
Fig. 3 shows that the template for the **Single Variable Prediction analytic** has been placed in the application workspace and the data from the **Bill and Tom** K Model is propagated into the two drop down boxes in the **Prediction** area. The first box contains the list of field names. The second box contains the list of field variables for the field currently selected in the first box.
There are several analytic templates in the **Analytics** menu list. This is an area of development and many other analytics are currently in development or proposed.

Sample Data Set – Sales Team Activities

Bill Tuesday 100 sold PA
Bill Tuesday 100 sold PA
Bill Tuesday 100 sold PA
Bill Tuesday 100 sold PA
Bill Tuesday 100 sold PA
Bill Tuesday 100 sold PA
Bill Monday 103 sold NJ
Bill Monday 100 trial PA
Bill Monday 100 trial PA
Bill Monday 100 trial PA
Tom Monday 100 sold PA
Tom Monday 100 sold PA
Tom Monday 103 trial NJ
Tom Monday 103 trial NJ
Tom Monday 103 trial NJ

Fig.2

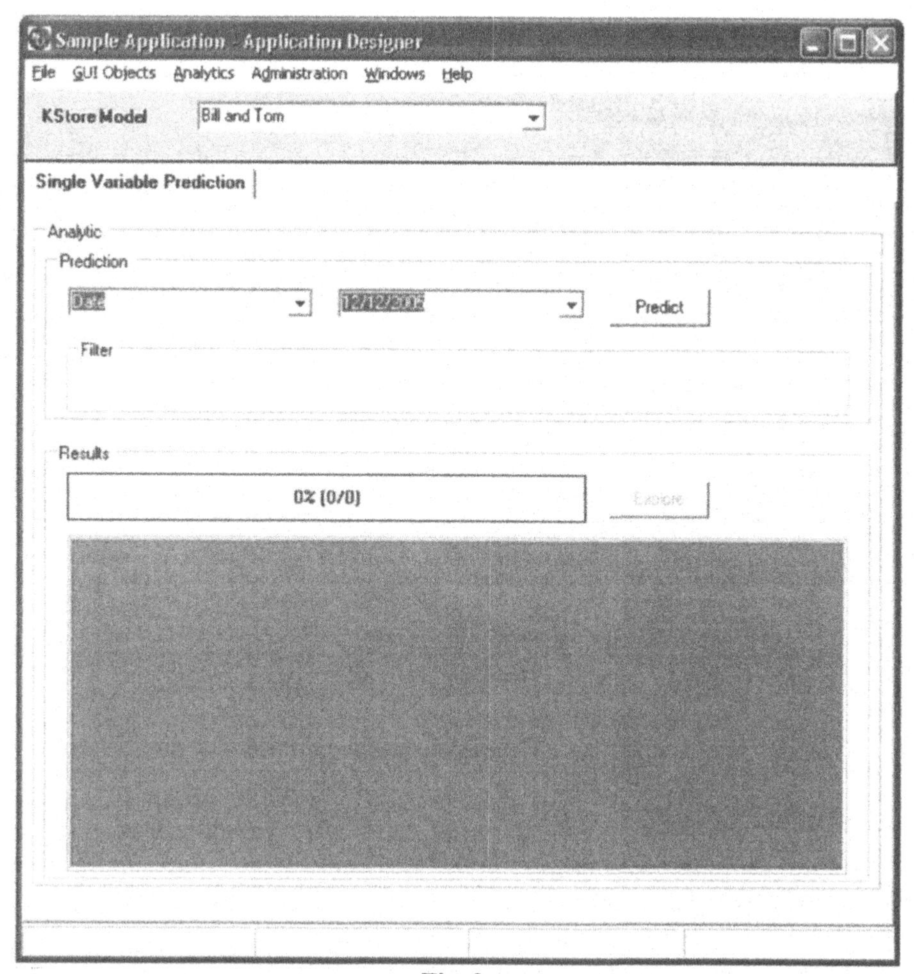

Fig. 3

After the application was created, the **Salesperson** field was selected from the first drop down box in the prediction area, and the field variable **Bill** was selected from the second drop down box in the prediction area. This indicates that the probability of **Salesperson/Bill** is requested.

In addition to specifying the field variable **Salesperson/Bill,** filters or constraints can be applied to the data set before the probability of **Salesperon/Bill** is calculated. In this example **Dayof Week/Monday** has been entered into the filter area.

The **Predict** and **Explore** buttons are clicked to run the calculation and display the record grid.

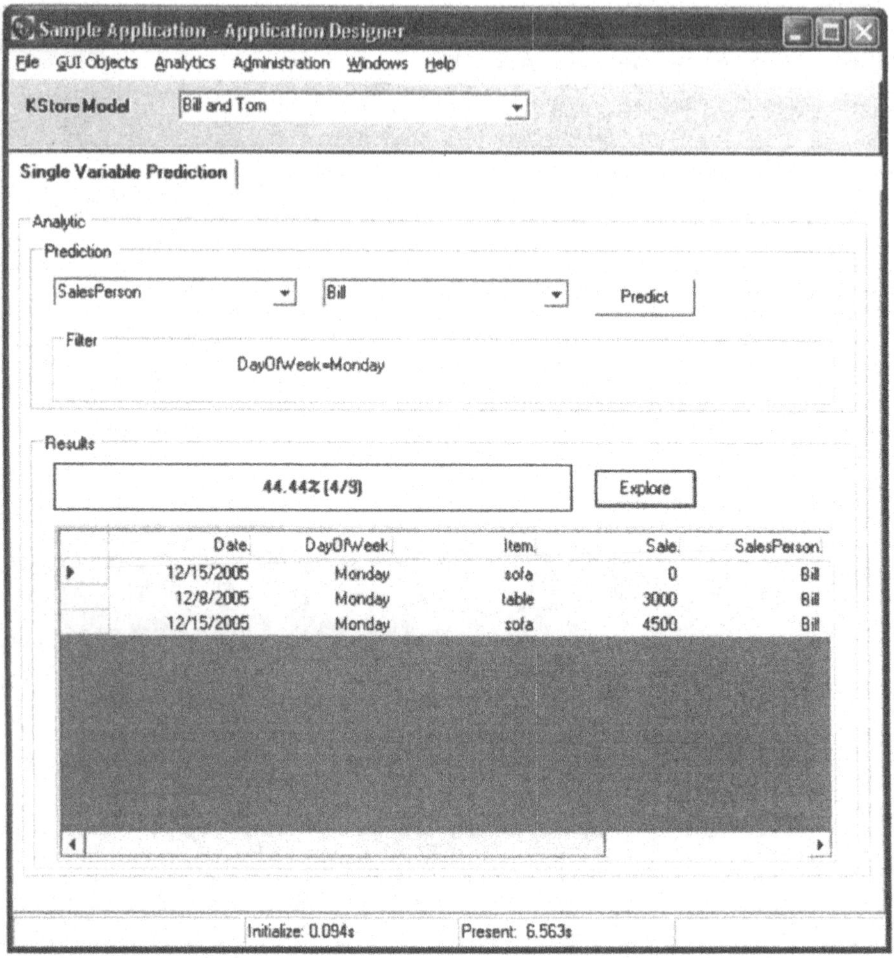

Fig.4

Fig. 4 shows the results for the **Salesperson/Bill** within the records that also contain **DayofWeek/Monday.** There are 9 records that contain **DayofWeek/Monday** and 4 of those records also contain **Salesperson/Bill.** So the probability of a record containing **Bill** is 4/9.

The grid shows each unique record containing **Salesperson/Bill and DayofWeek/Monday.**

4 Conclusion

The Application Designer has the tools required to create K Models and deploy GUI analytics to analyze them. These GUI applications are used primarily to demonstrate the flexibly and dynamic qualities of the K data structure, particularly in the analysis of large data sets. A K data structure models the relationships within a data set directly from raw data and can support any analytic without the need for data structures external to the K, such as cubes or indexes.

Applications that use K Models can also be created by GUI or batch softwares that call the API directly.

References

[KS] Mazzagatti, Jane C.: "A Computer Memory Resident Data Structure Based on the Phaneron of C. S. Peirce" in Inspiration and Application, Contributions to ICCS 2006 14th International Conference on Conceptual Structures ICCS 2006: Aalborg University Press 2006: Pascal Hitzler, Henrik Scharfe eds.: July 2006, pp. 114-130

[CP] Peirce, Charles S.: Collected Papers of Charles Sanders Peirce: 8 vols.: Vols. 1-6 ed. Charles Hartshorne and Paul Weiss: Vols. 7-8 ed. Arthur Burks: Cambridge: Harvard University Press, 1931-58

[PT] Patents
U.S. Patent Nos. 6,961,733 and 7,158,975 and patents pending

Mobile Transactional Modelling: From Concepts to Incremental Knowledge

Ivan Launders[1], Simon Polovina[2], and Richard Hill[2]

1 Mobility Solutions, BT Global Services, PO Box 200, London, N181 1ZF United Kingdom
ivan.launders@bt.com
2 Cultural, Communication and Computing Research Institute
Faculty of Arts, Computing, Engineering and Sciences, Sheffield Hallam University
City Campus, Howard Street, Sheffield, S1 1WB United Kingdom.
{s.polovina, r.hill}@shu.ac.uk

1 Introduction

The combination of ubiquitous mobile networks coupled with smart applications that are capable of working in highly dynamic environments will introduce a set of mobile transactional software design challenges. These design challenges require good analysis techniques and practices.
In an integrated system the same knowledge base that drives an expert system should also generate the specifications for conventional programs and databases that interact with it, according to Sowa [6]. Often an integrated system is constrained by the level of the transaction imposed through system boundaries and system interfaces. A mobile transaction offers the opportunity to remove some system boundaries and to re-model based on new mobile transaction goals.

2 Abstracting Mobile Transaction

Abstracting mobile transaction rules could lead to higher more abstract business goals being achieved by a system. If the mobile transactional rules are abstracted, then mobile transactions have the potential to become enriched. For example a health worker might be working with a specific health application providing access to a patient's records and prescriptions. Access to the application could be through a fixed network in a fixed location, constraining a clinician to access patient records through a specific application on a fixed network at a specific office location.
If the clinician's job role is mobile, i.e. a clinician visiting an elderly or immobile patient then the workflow of the clinician is constrained by the level of fixed transaction allowed by the existing systems. Once fixed system boundaries are removed, in this case the clinician is able to gain trusted access to a patient's records at their home through a GPRS or 3G link. It follows that the work pattern of that

mobile clinician has the potential to become more efficient. Trusted access to a patient's records means providing authentication of identity and evidence or the trustworthiness of data, services, and agents.

It may be possible for the clinician to adapt and learn new work patterns and to develop more efficient working practices if the mobile transactions provided by the system are designed to be as rich as possible. In this example removing the physical constraints through a mobile solution allows more abstract transaction goals to be aimed for. Transactions have the potential to become richer in terms of economic and care value between the patient and the clinician. There is a direct economic event associated with the transaction, in that the clinician is able to avoid needless travel to a trusted network access point in order to access patient records before visiting patients. There is an implied debit and credit to the mobile transaction, with the credit being increased through an improvement in work practice. Polovina [4] and Hill et al. [3] propose that transactions in a transaction framework recognise the consequential benefits and costs that occur through the exchange or resources.

Analysing the nature of a mobile transaction is important as it could allow for the identification of the higher value mobile transaction rules; those rules that provide the greatest benefit to a transaction for a given context. For example, the relevant circumstances for each visit a clinician makes will be different because the patient is likely to have a different need. The change in context or relevant circumstance means that it may be possible to improve the mobile transaction using different or learned knowledge from a given set of circumstances.

Categorising and modelling the concepts between mobile transaction rules and system boundaries could benefit the design of mobile solutions. The TrAM framework provides a model employing conceptual graphs to enrich the gathering of early requirements.

Conceptual graphs allow the analyst to use interactive tools for translating natural language into a knowledge representation language. The knowledge representation language can generate the following information according to Sowa [21]:

- Rules for driving an expert system;
- Specifications for conventional programs;
- Database definitions for the data dictionary;
- Help and diagnostic aids for programmers and end users.

Conceptual analysis enables a system's analyst to define new concepts schemata for a knowledge based system. TrAM provides a framework that builds upon the rigour of CGs by providing model checking to assist in the early requirements capture.

3 Applying Mobile Transaction Agent Modelling (TrAM)

The early analysis stage in Mobile Transaction Design is about understanding the problem in a mobile dimension and then defining the requirements and their goals. Identifying a model relevant to the domain and context, and then constructing a model to test the design is the challenge. Which models work and what kind of ontology can be constructed, depends on the nature of the mobility problem. The Ontology provides the means to classify things, to give classifications names and labels and to define the kinds of mobile properties and mobile relationships they can be assigned.

In general terms there are several steps in the analysis stage; the first step is understanding the requirements, the next step is the construction of classes, relations, functions and other objects which are part of the ontology and the final step is constructing models, for example a conceptual model. TrAM provides a framework which employs Conceptual Graphs to enrich the gathering of early requirements by:

a) Providing a means of capturing and modelling high-level, qualitative concepts;
b) Exploiting the formal underpinnings of TrAM's use of Conceptual Graphs and through the use of Peirce's logics, explained by Dau [2] to enable consistency checks in the notation to be made;
c) Using TrAM the Transaction Model (TM) to provide both design guidance and a mechanism for checking high-level transactions;
d) Deriving a hierarchy of types and a set of constraints upon which an ontology can be built.

The TrAM approach is as follows:

a) Capture high-level conceptual models demonstrating qualitative and political influences upon the case study;
b) Specialise the Transaction Models, according to the mobile domain illustrating the duality relationships between events and resources;
c) Identify an Enhanced Hierarchy of concept types and an audit trial of the key modelling decisions;
d) Ontology development from these requirements.

TrAM first captures some concepts, the principle being you don't need to be a subject area expert to capture them, according to the principle that it is an unconstrained view of the subject domain. Expert knowledge is then introduced whilst providing knowledge feedback, thus expert knowledge develops in light of checks and balances that the model provides. One level of expert knowledge will be different from another, the point being the process has a starting point.
E.g. Transform with TM, Gather use case, and Verify TM Graphs.

4 Case Study Example

In our experiences, students applying TrAM to case studies such as e-learning, health, and financial services have adopted the approach laid out in the following section. Some of the steps adopted are arbitrary whereas some are prescribed. The arbitrary steps seem to have been adopted to aid familiarisation and to understand the problem.

4.1 Transactional analysis – understanding the problem

Transaction analysis revolves around identifying those parts of the DFD that can be converted to transaction-centred structure charts.

1. Use case analysis and diagram – A use case describes the steps that people follow in carrying out a system activity. Often the first step is to identify the major activities and show them in terms of a use case model;
2. Draw the Conceptual Graphs (CG);

3. Identify the Transaction Modelling (TrAM);
4. Identify the Type Hierarchy (TH) – The type hierarchy supports the inheritance of properties from supertypes to subtypes of concepts. Studying the type hierarchies using the dictionary conceptual catalogue.
   a. Apply Peirce logic to the business rules;
   b. Refine the model and the Hierarchy showing how TrAM models map concepts to types;
5. Map TrAM to a UML class diagram
   a. Full class diagram;
   b. System boundary diagram – The components that make up the system, anything outside the system boundary is known as the system environment. The system boundary is the set of system components that can be changed during system design;
6. Analysis of Implementation related Issues.
   Clearly the above is an iterative process which has to be captured through TrAM.

5 Modelling Rules Using Incremental Knowledge

Ripple Down Rules (RDR) facilitates incremental knowledge acquisition, providing a framework for direct knowledge elicitation from experts after an initial setup by knowledge engineers. Rules are buried in various parts of knowledge-based or expert systems. An approach is to extract transactional based rules using TrAM and then to model those rules using RDR. In an RDR framework, the expert adds a new rule based on the context of an individual case.

Early versions of RDR were focused on classification tasks, firstly single classification demonstrated by Compton, Edwards et al. [1]. Richards and Compton [5] went onto investigate the use of formal concept analysis.

In a RDR knowledge base, the rules are stored in the tree. Each node of the tree is a rule. In the simplest version of RDR, each node has only two child-branches. One branch signifies that the conditions specified in the rules are satisfied, the other when they are not. Each node also has a conclusion that should be adopted if the conditions are satisfied.

RDR are flexible as it is not necessary to analyse the whole system before adding a new rule. Rules are only ever added to the ends of the tree, so they only effect the outcome of the rule above. By implementing a subset Conceptual Graph (CG), rules using RDR may be possible to model and categorise the high value rules in a mobile transaction.

An RDR inference engine interprets a case by evaluating it against some rules in a particular order. This ordering is imposed by the RDR tree structure. The principle of inference is to derive new knowledge from knowledge that we already know.

A classification from a rule is asserted to memory if the rule is fired by the current case. A rule is fired by a case if all conditions (or Boolean features) of this rule are satisfied.

Conclusion

The use of TrAM could provide a base in terms of being able to abstract structures and describe mobile transactions. The ability to model and represent incremental

knowledge in a mobile domain using concepts and the relationships between those concepts means expert knowledge can be captured and modelled as part of the design cycle. This ensures mobile transactions are designed to be as rich as possible.
Work focused around developing a technique using TrAM and RDR to model mobile transactions is of value as it allows the nature of a mobile transaction to be explored and represented in a design. The area of being able to abstract knowledge in a mobile environment and then to model mobile transactions is the subject of our ongoing work.

References

Compton, P., Edwards G., Kang B., Lazarus L., Malor R., Menzies T., Preston, P., Srinivasan, A., and Sammut, C., (1991) 'Ripple Down Rules: Possibilities and Limitations' 6th Banff Knowledge Acquisition for Knowledge-Based Systems Workshop, Canada, pp. 6.1-6.18.

Dau, Firthjof (2006) 'The Role of Existential Graphs in Peirce's Philosophy' In: Øhrstrøm, P.; Schärfe, H.; Hitzler, P.\ (Eds.): Conceptual Structures: Inspiration and Application: Contributions to ICCS 2006. (ISBN: 87-7307-768-2), Aalborg University Press, Aalborg, Denmark.

Hill, R., Polovina, S., Shadija, D., (2006) 'Transaction Agent Modelling: From Experts to Concepts to Multi-Agent Systems', Proceedings of 14th International Conference on Conceptual Structures (ICCS '06), Springer (ISBN 978-3-540-35893-0, ISSN 0302-9743), pp. 247-259.

Polovina, S., (1993) 'Bridging Accounting and Business Strategic Planning Using Conceptual Graphs', Proceedings of the Annual Conceptual Graphs Workshop, 1992, Springer-Verlag, Berlin (ISBN 3-540-57454-9), pp. 312-321.

Richards, D., and Compton, P., (1997) 'Uncovering the conceptual models in RDR KBS', International Conference on Conceptual Structures ICCS'97, Seattle, Springer-Verlag, pp. 198-212.

Sowa, John F., 'Conceptual structures: information processing in mind and machine', Addison-Wesley, 1984.